Compiling Esterel

Compiling Esterel

Dumitru Potop-Butucaru

Stephen A. Edwards

Gérard Berry

A C.I.P. Catalogue record for this book is available from the Library of Congress.

ISBN 978-0-387-70626-9 (HB)
ISBN 978-0-387-70628-3 (e-book)

Published by Springer,
P.O. Box 17, 3300 AA Dordrecht, The Netherlands.

www.springer.com

Printed on acid-free paper

All Rights Reserved
© 2007 Springer
No part of this work may be reproduced, stored in a retrieval system, or transmitted in any form or by any means, electronic, mechanical, photocopying, microfilming, recording or otherwise, without written permission from the Publisher, with the exception of any material supplied specifically for the purpose of being entered and executed on a computer system, for exclusive use by the purchaser of the work.

To Maria, Eleonora, and Victoria

Preface

This book is about compiling the Esterel language. One of us (Berry) created Esterel in the early 1980s for programming embedded reactive systems. Over time, we and others have developed both academic and commercial compilers for it that generate software and hardware. The newest version of Esterel, called Esterel V7, is being standardized by the IEEE as of 2006.

This book had its genesis in 2002 as Potop-Butucaru's thesis. Written under Berry and Robert de Simone, it made two main contributions: the first semantics for Esterel that included data manipulation (long a part of the language, this was its first formalization), and a very sophisticated code generation technique that remains one of the best developed so far.

At Potop-Butucaru's thesis defense, held during the SYNCHRON workshop in La Londes les Maures, France in November 2002*, Edwards, a jury member, remarked that the thesis would make a good book. Potop-Butucaru interpreted this as meaning only that the thesis would make a good starting point for the book. Thus, one of us (Edwards) expected to have a book within a few months, but instead it took years.

Little more than the general outline of the original thesis remains at this point; much has been added and clarified. We added chapters on the behavioral semantics of Esterel (Chapter 4), recently-developed compilation techniques used in the Columbia Esterel compiler (Chapter 10), and appendices on the extra constructs in the language that are not covered formally but must be handled by all compilers (Appendix A), the first formal language reference manual for the dialect of Esterel described in this book (Appendix B), the C language interface (Appendix C), and a description of the new Esterel V7 dialect of Esterel (Appendix D).

How to read this book

We divided this book into three parts. In the first part (Chapters 1–2), we provide an intuitive description of the Esterel language—enough to familiarize the reader with Esterel's synchronous model of time and the implications it

---

*A major side benefit of working on Esterel is that most of the meetings take place along the Côte d'Azur, which, not coincidentally, includes the Esterel region after which the language was named.

has for compilation. This should also be enough to get one started writing Esterel programs, but it does not presume to be a comprehensive tutorial on coding in Esterel.

The second part (Chapters 3–6) describes the formal semantics of Esterel, extending the work of Berry [7]. Why do we need three very different presentations of Esterel semantics? Each takes a different approach to describing how an Esterel program is to be executed and is appropriate for addressing different issues in code generation. The behavioral semantics (Chapter 4) is the simplest we present and most clearly addresses some of the thornier aspects of Esterel's rich control constructs and intra-cycle data dependencies. Unfortunately, it is a poor starting point for implementing the language as it represents control by rewriting the program. This makes for a convenient formalism but would make for an extremely inefficient implementation.

The practical shortcomings of the behavioral semantics lead us to the operational semantics (Chapter 5), which uses a more complicated formalism that represents program state by decorating the program text with all manner of diacritical marks. It is a much more complicated formalism as a result, but much closer to a practical implementation that represents program state as some sort of persistent marking of a program.

Unlike the behavioral and operational semantics, which use Plotkin's structural operational style, the third style presented in the second part—the circuit translation—shows how to translate an Esterel program into a circuit netlist. While this may appear surprising for a compiler that produces software, it turns out to be quite effective as the semantics of the circuit model very closely match those of Esterel and are much simpler. In fact, all the efficient compilation techniques that we present in the third part of the book start from GRC—a circuit-netlist-like intermediate representation.

Finally, the third part of the book (Chapters 7–10) gets at our main point: how to translate Esterel programs into efficient software implementations. After an overview, we present the GRC intermediate format (Chapter 8)—a representation developed by Potop-Butucaru as part of his thesis work that has become the foundation for a number of compilers. GRC is a hybrid representation that resembles both a traditional control-flow graph and a circuit netlist; it embodies all that we have learned about the structure of Esterel semantics in the past twenty years. Generating efficient code from it is natural; we describe the basic translation in Chapter 9. Finally, in Chapter 10, we describe how a slight variant of the representation has been used as the basis of two back ends in the open-source Columbia Esterel compiler.

The four appendices contain information that had previously been scattered in various technical reports and tutorials. Appendix A discusses parts of the language that are not usually included in the formal presentation of the language semantics, including valued signals, the `pre` operator, concurrent `trap-exit` abortion handling, and task control. Most amount to syntactic sugar, but deserve the more formal treatment we give them here.

Appendix B is the first semi-formal language reference manual for the V5 dialect of Esterel we use in this book. Previously, this information has been presented in a less formal tutorial style that was not as suitable for writing a compiler. See the Esterel V5 primer [8].

Appendix C presents the C language interface used by all the compilers described in this book. This interface was first used in the INRIA compiler and has since become the de facto standard for code generated from Esterel. The appendix explains, among other things, how to interface with the C code generated by any Esterel compiler and actually make use of it, which in some sense is the whole point of this book.

Finally, Appendix D presents the new Esterel V7 dialect, which has been under development at Esterel Technologies since 2001 as a broad extension of Esterel V5. The compilation techniques presented in this book are being applied to this dialect, as its core semantics remain the same. The Esterel V7 language is open (not proprietary) and its language reference manual [29] has been submitted for IEEE standardization.

Acknowledgements

Potop-Butucaru would like to acknowledge the supervision and help of Robert de Simone in writing of the original Ph.D. thesis.

Contents

Preface   **vii**

I   The Esterel Language

1   Introduction to Esterel   **3**
    1.1   Reactive Systems 3
    1.2   The Synchronous Hypothesis 4
    1.3   Implementation Issues 5
    1.4   Causality 6
    1.5   Related work 6
    1.6   A First Esterel Example 7
    1.7   Causality Cycles 8
    1.8   Code Generation 9
         1.8.1   Translation to Explicit State Machines 9
         1.8.2   Translation to Circuits 10
         1.8.3   Direct Compilation to C Code 12
    1.9   Executing the Generated Code 12
         1.9.1   Existing Solutions 13

2   The Esterel Language   **15**
    2.1   Syntax and Naïve Semantic Principles 15
    2.2   The Kernel Esterel Language 20
    2.3   Esterel Through Examples 24
    2.4   Host Language 31
    2.5   Program Structure and Interface 31
         2.5.1   Data Handling 32
         2.5.2   Signal and Signal Relation Declarations 34
         2.5.3   The **run** Pseudo-Statement 34

II Formal Semantics

3 Introduction to Esterel Semantics   41
   3.1 Intuition and Mathematical Foundations 41
          3.1.1 The Constructive Approach 43
   3.2 Flavors of Constructive Semantics 47
   3.3 Conventions and Preliminary Definitions 50
          3.3.1 Global Correctness of an Esterel Program 50
          3.3.2 Restriction to Kernel Esterel 51
          3.3.3 Signal Events 51
          3.3.4 Trap Handling and Completion Codes 51

4 Constructive Behavioral Semantics   55
   4.1 Behavioral Transitions 55
          4.1.1 Transition syntax 55
          4.1.2 States as Decorated Terms 56
          4.1.3 State Syntax 57
   4.2 Analysis of Potentials 58
          4.2.1 The Definition of *Must*, *Can*, and Can^+ 60
          4.2.2 Elementary Properties 68
   4.3 Semantic Rules 68
   4.4 Proof 72
   4.5 Determinism 72
   4.6 Loop-Safe Programs. Completion Code Potentials 73
   4.7 Program Behavior 75

5 Constructive Operational Semantics   79
   5.1 Microsteps 80
   5.2 COS Terms 80
          5.2.1 Control Flow Propagation 81
          5.2.2 State-Dependent Behavior 81
          5.2.3 Syntax of Semantic Terms 82
   5.3 Data Representation 84
   5.4 Semantic Rules 85
          5.4.1 Rules for Pure Esterel Primitives 85
          5.4.2 Rules for Data-Handling Primitives 92
   5.5 Analysis of Potentials 95
          5.5.1 Reduction to Non-Dotted Terms 96
          5.5.2 Non-Dotted Terms over Dataless Primitives 97
          5.5.3 Non-Dotted Terms over Data-Handling Primitives . . 99
   5.6 Behaviors as Sequences of Microsteps 99
   5.7 COS versus CBS 101

6	Constructive Circuit Translation	103
	6.1 Digital Circuits with Data	104
	6.1.1 Circuit Semantics. Constructive Causality	104
	6.1.2 Extension to Circuits with Data	107
	6.1.3 Formal Definitions	110
	6.2 Translation Principles	112
	6.2.1 The Selection Circuit	114
	6.2.2 The Surface and Depth Circuits	115
	6.2.3 The Global Context	116
	6.3 Translation Rules	117
	6.3.1 Dataless Primitives	117
	6.3.2 Data-Handling Primitives	126
	6.4 Circuit Translation versus COS	128

III Compiling Esterel

7	Overview	135
	7.1 Compiler Classes	135
	7.2 A Brief History	136
	7.3 The INRIA Compiler	137
	7.4 The Synopsys Compiler	139
	7.5 The Saxo-RT Compiler	140
	7.6 The Columbia Esterel Compiler	144

8	The GRC Intermediate Format	145
	8.1 Definition and Intuitive Semantics	146
	8.1.1 The Hierarchical State Representation	146
	8.1.2 The Control/Data Flowgraph	149
	8.1.3 Implementation Issues	154
	8.2 Esterel to GRC Translation	158
	8.2.1 Translation Principles	158
	8.2.2 Translation Rules	159
	8.2.3 The Global Context	167
	8.3 Formal Simulation Semantics and Translation Correctness .	167
	8.4 Format Optimizations	168
	8.4.1 State Representation Analysis	170
	8.4.2 Flowgraph Optimizations	173

9	Code Generation from GRC	179
	9.1 Defining "Acyclic"	180
	9.2 Code Generation for Acyclic Specifications	185
	9.2.1 State Encoding	185
	9.2.2 Flowgraph Transformations	190
	9.2.3 Scheduling	192

9.3		Code Generation for Cyclic Specifications	193
9.4		Benchmarks	200

10 The Columbia Compiler   203
10.1 The Dynamic Technique 203
    10.1.1 An Example 204
    10.1.2 Sequential Code Generation 211
    10.1.3 The Clustering Algorithm 214
10.2 The Program Dependence Graph Approach 215
    10.2.1 Program Dependence Graphs 216
    10.2.2 Scheduling 218
    10.2.3 Restructuring the PDG 221
    10.2.4 Generating Sequential Code 228
10.3 Benchmarks 229

A Language Extensions   235
A.1 Signal Expressions 235
    A.1.1 Syntactic Aspects and Limitations 236
    A.1.2 Combinational Expressions 236
    A.1.3 The `pre` Operator 238
    A.1.4 Delay Expressions. Preemption Triggers 241
A.2 Traps and Trap Expressions 242
    A.2.1 Concurrent Traps and Trap Expressions 243
    A.2.2 Valued Traps 244
A.3 The `finalize` Statement 245
A.4 Tasks 247
    A.4.1 Task Synchronization Semantics 248
    A.4.2 Multiple `exec` 251

B An Esterel Reference Manual   253
B.1 Lexical Conventions 253
    B.1.1 Tokens 253
    B.1.2 Comments 253
    B.1.3 Identifiers 253
    B.1.4 Reserved Words 254
    B.1.5 Literals 254
B.2 Namespaces and Predefined Objects 255
    B.2.1 Signals and Sensors 256
    B.2.2 Variables and Constants 256
    B.2.3 Traps 256
    B.2.4 Types 257
    B.2.5 Functions and Procedures 257
    B.2.6 Tasks 257
B.3 Expressions 258
    B.3.1 Data Expressions 258

		B.3.2	Constant Atoms	258
		B.3.3	Signal Expressions	259
		B.3.4	Delay Expressions	259
		B.3.5	Trap Expressions	259
	B.4	Statements		260
		B.4.1	Control Flow Operators	260
		B.4.2	`abort`: Strong Preemption	261
		B.4.3	`await`: Strong Preemption	262
		B.4.4	`call`: Procedure Call	264
		B.4.5	`do-upto`: Conditional Iteration (deprecated)	264
		B.4.6	`do-watching`: Strong Preemption (deprecated)	264
		B.4.7	`emit`: Signal Emission	264
		B.4.8	`every-do`: Conditional Iteration	265
		B.4.9	`exec`: Task Execution	265
		B.4.10	`exit`: Trap Exit	266
		B.4.11	`halt`: Wait Forever	266
		B.4.12	`if`: Conditional for Data	266
		B.4.13	`loop`: Infinite Loop	267
		B.4.14	`loop-each`: Conditional Iteration	267
		B.4.15	`nothing`: No Operation	268
		B.4.16	`pause`: Unit Delay	268
		B.4.17	`present`: Conditional for Signals	268
		B.4.18	`repeat`: Iterate a Fixed Number of Times	269
		B.4.19	`run`: Module Instantiation	270
		B.4.20	`signal`: Local Signal Declaration	271
		B.4.21	`suspend`: Preemption with State Freeze	272
		B.4.22	`sustain`: Emit a Signal Indefinitely	273
		B.4.23	`trap`: Trap Declaration and Handling	273
		B.4.24	`var`: Local Variable Declaration	273
		B.4.25	`weak abort`: Weak Preemption	274
	B.5	Modules		275
		B.5.1	Interface Declarations	276
C	The C Language Interface			**281**
	C.1	Overview		281
	C.2	C Code for Data Handling		283
		C.2.1	Defining Data-handling Objects	283
		C.2.2	Predefined Types	283
		C.2.3	User-defined Types	283
		C.2.4	Constants	286
		C.2.5	Functions	286
		C.2.6	Procedures	287
	C.3	The Reaction Interface		288
		C.3.1	Input Signals	288
		C.3.2	Return Signals	289

	C.3.3	Output Signals	289
	C.3.4	Inputoutput Signals	289
	C.3.5	Sensors	290
	C.3.6	Reaction and Reset	290
	C.3.7	Notes	291
C.4	Task Handling		292
	C.4.1	The Low-level Layer: ExecStatus	292
	C.4.2	The Functional Interface to Tasks	295

D Esterel V7 — 297

D.1	Data Support		298
	D.1.1	Basic Data Types	298
	D.1.2	Arrays	298
	D.1.3	Generic Types	299
	D.1.4	Bitvectors	299
	D.1.5	From Numbers to Bitvectors and Back	300
	D.1.6	Data Units	300
D.2	Signals		301
	D.2.1	Value-only Signals	301
	D.2.2	Temporary Signals	301
	D.2.3	Registered Signals	301
	D.2.4	Signal Initialization	302
	D.2.5	Oracles	302
D.3	Interfaces		303
	D.3.1	Interface Declaration	303
	D.3.2	Interfaces and Modules	303
	D.3.3	Mirroring an Interface	304
	D.3.4	Interface Refinement in Modules	304
D.4	Statements		304
	D.4.1	Expressions and Tests	304
	D.4.2	Static Replication	304
	D.4.3	Enhanced Emit and Sustain Statements	305
	D.4.4	Explicit and Implicit Assertions	306
	D.4.5	Weak Suspension	307
	D.4.6	Signal Connection by Module Instantiation	309
D.5	Multiclock Design		310
	D.5.1	Clocks and Multiple Units	311
	D.5.2	Simulation of Multiclock Designs by Single-clocked Designs	312

Bibliography — 315

Index — 323

Figures

1.1	Execution cycle	5
1.2	A possible execution trace for ABRO	8
1.3	The automaton and explicit FSM code for the ABRO example	10
1.4	Synchronous circuit model	11
1.5	Circuit code for ABRO (the reaction function)	11
1.6	Control-flow (pseudo-)code for ABRO	12
1.7	A possible asynchronous run of ABRO	13
2.1	An execution trace of `MainExample`	25
2.2	Interconnections between `Cell` modules in `Arbiter2`	28
2.3	A 3-cell FIFO in Esterel.	35
3.1	A correct, but complex Esterel example	43
3.2	Boolean circuits for two intuitionistic formulas	46
3.3	Truth table for ternary logic ($\mathcal{B}_\perp$) operators	46
3.4	Completion codes in the start instant of a simple statement	53
6.1	A possible circuit translation for ABRO	105
6.2	The two possible evaluation sequences for a small circuit	106
6.3	Causality in circuit evaluation	106
6.4	A simple sequential circuit	107
6.5	Circuit with data, first example	108
6.6	Circuit with data, second example	109
6.7	Circuit with data, third example	109
6.8	Incorrect circuits	112
6.9	The selection circuit of a simple example	115
6.10	The interface of the generated circuits	115
6.11	The global translation context	116
6.12	Surface and depth circuits for `pause`	117
6.13	Surface circuit for `nothing`	117
6.14	Surface circuit for "`loop` p `end`"	118
6.15	Depth circuit for "`loop` p `end`"	118

6.16	Surface circuit for the two-way sequence "$p\,;q$"	119
6.17	Depth circuit for the binary sequence "$p\,;q$"	119
6.18	Surface circuit for "$p_1 \parallel \ldots \parallel p_n$"	120
6.19	Depth circuit for "$p_1 \parallel \ldots \parallel p_n$"	120
6.20	The circuit-level parallel synchronizer	121
6.21	Surface circuit for "emit S"	121
6.22	Surface and depth circuits for "signal S in p end"	122
6.23	Surface circuit for the signal and data tests	122
6.24	Depth circuit for the test statements	123
6.25	Signal test circuit (present statement)	123
6.26	Surface circuit for "suspend p when $expr$"	124
6.27	Depth circuit for "suspend p when $expr$"	124
6.28	Surface circuit for "exit T(i)"	124
6.29	Surface circuit for "trap T in p end"	125
6.30	Depth circuit for "trap T in p end"	125
6.31	Surface circuit for the variable assignments	126
6.32	Surface circuit for variable declarations	127
6.33	Depth circuit for variable declarations	128
6.34	Variable test circuit (if statement)	128
7.1	Esterel implementation flavors	136
7.2	The flow of the INRIA compiler	138
7.3	Esterel fragment and its Synopsys translation into C	140
7.4	Concurrent control-flow graph, Synopsys style	141
7.5	A simple Esterel program modeling a shared resource	142
7.6	Event graph and C code generated by the Saxo-RT compiler	143
8.1	The MainExample Esterel program	147
8.2	The selection tree of MainExample	148
8.3	Selection tree and selection flags for a small example	148
8.4	Simple GRC flowgraph	149
8.5	Signal dependency representations	150
8.6	The flowgraph of MainExample	153
8.7	The Esterel to GRC translation interface	158
8.8	The translation of pause	159
8.9	The translation of nothing and emit S	160
8.10	The translation of loop q end	160
8.11	The translation of $q\,;r$	161
8.12	The translation of $q \parallel r$	162
8.13	The translation of signal S in q end	162
8.14	The translation of the test primitives	163
8.15	The translation of suspend q when $expr$	164
8.16	The translation of exit $T(i)$	164
8.17	The translation of trap T in q end	165

8.18	The translation of variable assignment	165
8.19	The translation of variable declarations	166
8.20	The global translation context	166
8.21	Circuit counterpart for the **Test** node	167
8.22	Circuit counterpart for the **Switch** node	168
8.23	Tagged tree for `MainExample`	173
8.24	Simplified graph for `MainExample`	177
9.1	A GRC-level cycle resolved at the circuit level	181
9.2	Simplified translation patterns for `present`	182
9.3	Example of synchronizer refinement	184
9.4	Bit allocation for our first small example	186
9.5	State encoding examples for our first small example	187
9.6	Bit allocation for `MainExample`	188
9.7	State encoding examples for `MainExample`	188
9.8	Bit allocation for the third small example	189
9.9	State encoding examples for the third small example	189
9.10	Bit allocation for `pause ; pause`	191
9.11	Possible states for `pause ; pause` and their encoding	191
9.12	Flowgraph transformation due to state encoding	192
9.13	A partition of the selection tree of `MainExample`	193
9.14	Code generated from the flowgraph of Figure 8.24	194
9.15	The GRC of the cyclic example	196
9.16	The SCC at GRC and circuit level	196
9.17	Globally acyclic graph for our cyclic example	197
10.1	An Esterel model of a shared resource	205
10.2	The GRC_{CEC} for the program in Figure 10.1	206
10.3	The control-flow graph from Figure 10.2 divided into blocks	209
10.4	The code CEC generates for part of the graph in Figure 10.3	210
10.5	Cluster code and the linked list pointers	213
10.6	The clustering algorithm	214
10.7	A program dependence graph requiring interleaving	217
10.8	The Main procedure	219
10.9	Successor Priority Assignment	220
10.10	The Scheduling Procedure	221
10.11	The Restructure procedure	221
10.12	The DuplicationSet function	222
10.13	The DuplicateNode procedure	223
10.14	The ConnectPredecessors procedure	223
10.15	The restructured PDG from Figure 10.7	226
10.16	A complex example	227
10.17	The reconstructed PDG from Figure 10.7	229

10.18	The PDG of Figure 10.17 after guard variable fusion	231
10.19	The successor ordering procedure	232
A.1	Test sub-circuit for the expression "`C and not (A or B or C)`"	237
A.2	Tagged selection tree for the expansion of `pre`	239
A.3	The life cycle of a task	249

Tables

2.1	The primitives of Pure Esterel	22
8.1	Parallel branch status redundancy statistics	172
9.1	State encoding results on some typical examples	190
9.2	Code speed	202
9.3	Code size	202
10.1	Experimental results for the dynamic approach	230
10.2	Statistics for the examples	230
10.3	Experimental Results for the PDG-based approach	233

Part I

The Esterel Language

1
Introduction to Esterel

Esterel is a synchronous programming language tailored for the development of control-dominated embedded reactive applications in both hardware and software. In this chapter, we explain what this means. We explain what a reactive system is, introduce the key concepts of synchronous reactive system programming, and give a short overview of the Esterel language and framework.

1.1 Reactive Systems

Reactive systems, defined in the 1980s [41, 40, 39], are computer systems that continuously react to input events from their environment by producing appropriate output events in a timely manner. They differ from *transformational systems*, which emphasize data computation instead of system-environment interaction; and from *interactive systems*, which react to environment requests at their own rate instead of at the rate required by the environment. For instance, an airplane autopilot is reactive while a web browser is interactive. Classical programming and verification techniques were originally geared to transformational systems and later extended to interactive systems. Unfortunately, they turn out to be inadequate for reactive systems. It is just too much of a stretch.

To address this problem, several groups in the 1980s developed specific techniques geared to the design and verification of reactive systems. One technique, the synchronous approach, is based on mathematical semantics and has generated a variety of languages and compilers that are now widely used in applications ranging from safety-critical embedded systems in avionics, automotive, and railways to electronic circuit design.

Reactive systems have two essential traits: concurrency and determinism. They are invariably composed of concurrent components that react concurrently to their environment. While interactive systems may also have this structure, there is a fundamental philosophical difference between the two:

The concurrent pieces in a reactive system are mostly concerned with cooperating and communicating in a deterministic way. The concurrency in an interactive system usually involves much more competition over shared resources, which is resolved in a nondeterministic way. Such nondeterminism renders classical concurrent programming techniques such as the shared memory and locks model inadequate for reactive systems.

The synchronous approach to reactive systems, embodied in the Esterel, Lustre, and Signal languages [39, 3, 4, 58] is among the solutions that have been proposed for the development of reactive systems. This concurrent model provides determinism and convenient mechanisms for cooperation. Furthermore, its crisp formal semantics simplifies verification. The Esterel synchronous language is the focus of this book.

Reactive systems are often embedded in objects that interact with the physical world, e.g., airplanes or automotive controllers, and their correct function is very often safety-critical. A system crash in an airplane is a much more literal problem than one on a desktop computer. Thus, validation and certification of reactive systems is an absolute must. Experience has shown that the synchronous approach can greatly improve the verification process, in particular by being well-suited to formal verification.

Reactive systems are also made of heterogeneous hardware and software components. A further advantage of the the synchronous approach, and of Esterel in particular, is that it applies equally to hardware and software. The Esterel V7 production compiler generates C/C++/SystemC code as well as VHDL/Verilog code with exactly the same behavior. The same verification algorithms apply to both. This does not solve the issue of hardware/software partitioning, but is an enabler for it: when writing a module, one does not need to decide upfront whether it will be implemented in hardware or software. One can use a unique programming style instead of two widely different ones (e.g., C and register-transfer-level Verilog).

1.2 The Synchronous Hypothesis

The synchronous hypothesis, the fundamental principle on which Esterel is based, states that a system reacts to environmental events in no time. Furthermore, communication among system components is also done instantaneously. The synchronous hypothesis thus separates the notion of physical time, which arises from the physics of the environment, from the execution time of the system, which is largely a side-effect of how it is implemented. An Esterel specification only refers to physical time, and assumes implementation time is exactly zero. This is a brutal but effective way to separate concerns, simplify the semantics, and to reconcile concurrency and determinism. It is the interference between physical and implementation time that makes classical asynchronous concurrent programs nondeterministic, hard to analyze, and difficult to write. For more on this philosophy, see Berry [9]. When

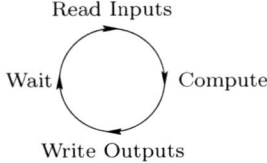

Figure 1.1: The execution cycle of the synchronous model

programming in a synchronous way, one concentrates on the functionality, postponing implementation details. Of course, one must check that the synchronous hypothesis is "correctly approximated" by the final implementation. We later show how this can be done.

Esterel relies on another fundamental principle: the discretization of time. The synchronous hypothesis as stated above applies equally well to continuous- and discrete-time systems. However, computer applications are most ofen based on discrete time. The discretization of time is ubiquitous in control and signal theory. It also applies to newer systems such as man-machine interfaces (e.g., cockpit displays), where the environment sends discrete events. Therefore, Esterel adopts a discrete model of time, where programs only react at discrete instants. At each reaction instant, a program reads its inputs, computes the outputs, and sends them to the environment, conceptually in no time.

1.3 Implementation Issues

Esterel is not just a specification language; it is meant to be implemented and used in actual systems. The object of this book is to present practical techniques for implementing theoretical zero-delay behavior, which is not obvious since any real computation takes time. The central idea is to approximate zero-delay reactions using the cycle-based model of computation pictured in Figure 1.1.

At any instant where the program should react, we perform a computation cycle: we sample the input, compute the output, and deliver it to the environment. Each of these operations takes some time. If the cycle computation is over before the next cycle should start, we have avoided input/computation interference exactly as required by the zero-delay model, and we can consider the computation as correct in practice.

Think of people talking to each other in a room. They can (and do) neglect the speed of sound, because the room is small enough to make communication practically instantaneous. By analogy, an Esterel implementation has two goals.

Speed. Efficiently fit the "people" inside the "room." That is, schedule the various computing elements of the system so that they perform their computations and communications fast.

Timing predictability. Make sure that the resulting system behaves like a "small room" where the communication speed can be neglected, so that reactions are predictable, and meet their timing constraints.

This book concentrates on speed issues; we do not address the important issue of predictability, which involves many non-trivial techniques such as worst-case execution time analysis (WCET), which can be performed using abstract interpretation and precise processor models. See for instance [42] for WCET analysis applicable to synchronous programs.

To respect the semantics, the execution of a cycle should be atomic. In an implementation, the environment should not be allowed to change during the cycle, and cycles should not overlap. Therefore, computations are performed on a frozen input snapshot taken during the input phase.

1.4 Causality

Causality among events is the key link between the zero delay model and a practical implementation. Roughly, an abstract event A causes an event B if the occurrence or non-occurrence of A determines the occurrence or non-occurrence of B. In the zero-delay model, events are determined by causal chain reactions among concurrent system components. In implementations, events are determined by sequences of computations that implement such causal reaction chains. Thus, compiling Esterel amounts to finding a way to implement causality chains with instruction sequences. This can be done in many ways, leading to the variety of implementation techniques studied here.

Unfortunately, not all synchronous programs are automatically causal. For instance, the logical contradiction "emit A if and only if A is not emitted" can be expressed as a synchronous program that is syntactically correct but has no synchronous semantics (note the similarity with the classical liar paradox). The theory of causal and non-causal programs is not a simple one; it is the subject of a separate book [7]. In practice, non-causal programs should be rejected by compilers with appropriate error messages. Most compilers take a conservative approach, sometimes rejecting projects that could be viewed as causal, but the conservative approach appears adequate in practice. In this book, we formally describe causality requirements and discuss which rejection algorithms are suitable for practical applications.

1.5 Related work

Esterel is only one of the synchronous languages; Argos [49], Lustre/Scade [38, 22], Signal [46], SyncCharts [1], and Quartz [62] also take the synchronous approach and benefit from the simplicity of the synchronous hypothesis. A key difference among these languages is the class of applications they target. Among these others, SCADE ("Safety-Critical Application Development Environment"), is perhaps the most successful. It is a graphical version of Lustre dedicated to safety-critical embedded software systems, and especially to certified avionics systems: flight control, cockpit display, engine control, brake control, etc. The SCADE compiler itself has been certified by avionics authorities, which greatly simplifies the laborious process of certifying the avionics software [24]. The simplicity and rigor of synchronous semantics and compiling techniques is the key to such a compiler certification.

Similar abstractions have lead to the development of quasi-synchronous formalisms such as Statecharts [40] or the synthesizable subset of the hardware description languages Verilog [44] and VHDL [51].

In compiler technology, the static single-assignment (SSA) intermediate representation [23, 45] requires that each variable has a single, non-recursive definition, so that it has a unique value throughout the execution. This is similar to synchronous causality, and facilitates various analysis and code generation techniques.

Synchrony is also commonly used for hardware design. In the Register Transfer Level (RTL) model, one describes circuits using gates that conceptually compute in zero time. After mapping the gates to hardware components, placing these components in space and routing the interconnect wires, one can statically analyze the timing of the circuit, i.e., compute the maximal time it takes for signals to stabilize. Then, the circuit behaves as in the zero-delay model if the period of the circuit clock is larger than the maximal stabilization time.* Unfortunately, although it is implicit in the definition of the synthesizable subset of VHDL [51], the zero-delay hardware model and its precise relation with timed models have not yet been formalized; the semantics of hardware description languages such as Verilog or VHDL remain informal.

1.6 A First Esterel Example

ABRO is perhaps the simplest interesting Esterel program. This program has three inputs, A, B, and R, and one output, O. It waits for the last of A and B to occur and then immediately emits O. The entire process restarts when R occurs. If R occurs at the same time as A or B, the reset takes priority. The Esterel code is as follows.

```
module ABRO:
input A, B, R ;
output O ;
loop
```

---

*If we refer to our "small room" analogy, such static timing analysis determines the room size that allows us to neglect communication time.

8 INTRODUCTION TO ESTEREL

Instant	Inputs	Outputs	Comment
0			
1	A,B	O	
2	A		Signal A ignored
3	R		Reset
4	A		Still waiting for B
5	B	O	
6	A,B,R		Reset, A and B ignored

Figure 1.2: A possible execution trace for ABRO

```
[ await A || await B ];
  emit O
each R
end module
```

This simple example features the four main ingredients of Esterel: signalling, concurrency, and preemption. It is actually a pattern commonly found in reactive designs: think of ABRO as a memory write controller, with A the address, B the value, O the write command, an R a synchronous reset.

Figure 1.2 presents a possible execution of ABRO. This trace is an indexed sequence of execution instants. At each instant, the input event tells which input signals have arrived, and the reaction determines whether to emit the output signal O. Note that input signals can arrive simultaneously.

At instant 0, no input arrives and no output is produced. The control enters the program, forks at the parallel statement, and is blocked at the concurrent await statements. At instant 1, the arrival of A and B makes both await statements terminate. Then, their parallel composition immediately terminates and passes control to the next statement in sequence: the statement emit O. This is immediately executed and outputs O. Then, control blocks, awaiting R. At instant 2, the input A is simply ignored because nothing is observing it. At instant 3, signal R arrives, the loop body is restarted, and ABRO waits again for A and B. When A arrives at instant 4, "await A" terminates. The parallel statement does not yet terminate; it is waiting for its right branch to terminate. This occurs at instant 5 because B arrives. O is immediately emitted. At instant 6, A, B, and R arrive at the same time. Then the "loop...each R" statement preempts is body and performs the reset, ignoring A and B.

1.7 Causality Cycles

Unfortunately, it appears the simplicity of synchrony comes at the price of the thorny problem of *causality cycles*. Reactions are computed by chains of instantaneous elementary actions, such as signal reception, statement termination, parallel synchronization, and signal emission. Control transmission makes these actions belong to causal chains. The problem is that the syntax of Esterel makes it possible to write cyclic causal chains that have no obvious meaning. For instance, it is easy to introduce a causality cycle to to ABRO example:

```
module Cycle :
input I;
output O;
signal A in
  run ABRO
||
  await O; emit A
end signal
end module
```

In Cycle, we remove A from inputs and declare it as a scoped local signal. The run statements instantiates ABRO, automatically connecting its inputs and outputs by name. The second parallel branch waits for O from ABRO to instantaneously feed A back to ABRO. The causality cycle is as follows: within ABRO, emission of O depends on reception of A; within the second parallel branch, emission of A depends on reception of O; both dependencies are instantaneous. The program makes no sense and should be rejected. The classical solution to avoiding causality cycles is to cut any instantaneous cycle by inserting a delay. This condition is indeed sufficient and useful in practice, but we will also study much finer conditions that have also proven useful. Formal semantics are essential to understanding the causality problem and for developing algorithms to reject non-causal programs with error messages that point to the root cause of the problem.

1.8 Code Generation

Three main techniques have been developed to translate Esterel programs into hardware or software: expansion to explicit state machines, translation into circuits, and direct software generation. In this book we shall focus on the generation of software. We shall not cover distributed software generation, studied for instance by Girault et al. [19], and we will address issues related to hardware generation only when they are relevant to generating software.

1.8.1 *Translation to Explicit State Machines*

The first INRIA compilers Esterel V2 (1985) and Esterel V3 (1988) translated Esterel programs into explicit finite automata, specifically Mealy state machines. These compilers explicitly enumerate all the reachable states and

10   INTRODUCTION TO ESTEREL

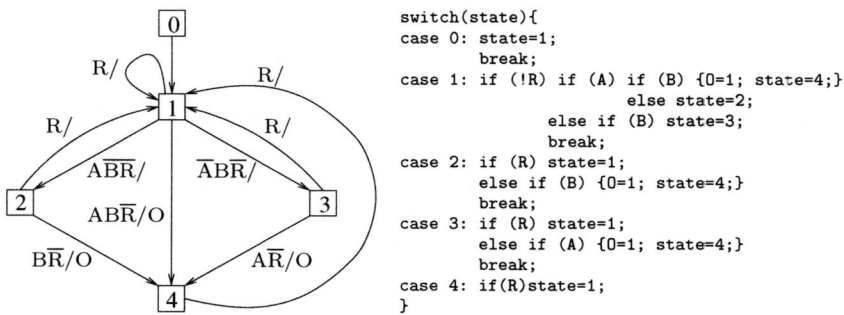

Figure 1.3: The automaton and explicit FSM code for the ABRO example

transitions of a program. This state machine is then encoded into C. We call this form of code *(explicit) FSM code*.

Figure 1.3 shows the state graph of ABRO along with its implementation in C. Notice the number of times A, B, R, and O appear in the state graph. With its constructs for concurrency and preemption, Esterel can be much more succinct than explicit finite automata for the same behavior. This is an essential advantage for clean and maintainable specifications.

The FSM code is very fast, since only the active code is evaluated in each instant. However, FSM translation is subject to exponential state space explosion.* The method can only be applied to small- and medium-sized programs.

Esterel V2 is based on Brzozowski's residual technique [17] to translate regular expressions into automata. Transitions are computed using a direct interpretation of the Esterel semantic rules, given in Plotkin's Structural Operational Semantics (SOS) style [5], and states were represented by textual Esterel programs. Causality analysis was simplistic and limited. Esterel V3 uses a more elaborate operational semantics due to Gonthier and Berry [10, 35], a much finer causality analysis that accepts many more programs, and an efficient bit-vector representation of states. Compilation is orders of magnitude more efficient in time and space than with Esterel V2, but the generated code is about the same and still subject to state explosion.

1.8.2   Translation to Circuits

The INRIA Esterel V4–V6 compilers (1992–1999) and the Esterel Studio commercial compiler use a very different technique: direct translation of an Esterel program into a digital circuit, developed by Berry [6, 7].

Figure 1.4 shows the standard form of a synchronous circuit. It is divided

---

*The FSM for ABRO has 5 states. Adding "await C" in the parallel construct produces 9 states. Further adding "await D" produces 17 states. In general, waiting for n signals in an ABRO-like program leads to an FSM with $2^n + 1$ states.

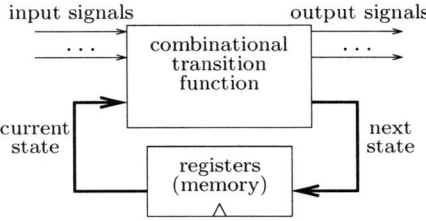

Figure 1.4: The standard representation of sequential circuits

```
RESET = R & !BOOT
A_TEST_TRIGGER = A_REG_OUT & !RESET
A_TEST_THEN = A_TEST_TRIGGER & A
A_TEST_ELSE = A_TEST_TRIGGER & !A
A_TERMINATED = A_TEST_THEN | ! A_TEST_TRIGGER
A_REG_IN = BOOT | RESET | A_TEST_ELSE
B_TEST_TRIGGER = B_REG_OUT & !RESET
B_TEST_THEN = B_TEST_TRIGGER & B
B_TEST_ELSE = B_TEST_TRIGGER & !B
B_TERMINATED = B_TEST_THEN | ! B_TEST_TRIGGER
B_REG_IN = BOOT | RESET | B_TEST_ELSE
ACTIVE = A_TEST_TRIGGER | B_TEST_TRIGGER
O = A_TERMINATED & B_TERMINATED & ACTIVE
```

Figure 1.5: Circuit code for ABRO (the reaction function)

into combinational logic, which computes the outputs and new states from the inputs and current state, and sequential logic (registers), which stores the state. A global clock ticks off the instants by triggering state updates. The circuit for ABRO is pictured in Figure 6.1, page 105.

The circuit translation technique scales to large programs since state is held in registers and encoded compactly, hence eliminating the exponential code-size explosion.

The circuit can be translated into VHDL or Verilog for implementation in hardware. After applying sequential optimization techniques developed by Toma et al. [69, 63] to improve the size and speed of the circuit, one gets very efficient circuits that are often better than manually designed ones. Sequential optimization does not scale to big programs, but can be applied on a per-module basis and produce reasonable results.

Software code is generated by topologically sorting the circuit gates and translating their function into C (or C-based languages such as C++ and SystemC), as pictured in Figure 1.5. We call this form of generated code *circuit code*. Compared to automaton code, circuit code is slower because in every instant it executes every equation, even those corresponding to inactive states.

The causality of the generated circuit is now well-studied. That the

```
ABRO_start() { //called in the initial simulation instant
    await_A_active=true ; await_B_active=true ;
}

ABRO_resume() { //called when ABRO is resumed
    if (R) { ABRO_start() ; }
    else {
      if (await_A_active || await_B_active) {
        if (await_A_active && A) { await_A_active=false ; }
        if (await_B_active && B) { await_B_active=false ; }
        if ((!await_A_active)&&(!await_B_active)) { emit O ; }
      }
    }
}
```

Figure 1.6: Control-flow (pseudo-)code for ABRO

combinational logic be acyclic is the simplest criterion, but such cyclic circuits may also be causal if every cycle is effectively broken at each instant by input or current state values. Cyclic circuits have been studied by Malik [48] and Berry and Shiple [65]. The latter showed that *constructive logic* can be used to model circuits with cycles. Such constructive analysis of circuits was later extended to the *constructive semantics* of the Esterel language itself [7] and implemented in the Esterel V5 compiler. The constructive semantics of Esterel is the subject of the second part of this book.

1.8.3 Direct Compilation to C Code

To improve upon circuit code, Edwards [26, 27], Closse and Weil [21, 72], and Potop-Butucaru [57, 56] developed new compilers based on control flow graph construction and optimization. These techniques will be explained in detail in this book. The commercial Esterel V7 commercial compiler embodies a C code generator based on Potop-Butucaru's technique.

The main idea in these compilers is to generate C code that compute only the active part of the Esterel program in each reaction, as does FSM code, while representing the state in a compact bit vector form, as done in circuit code. We call this form of generated code *control-flow code*. Figure 1.6 shows control-flow code similar to what Edwards's and Potop-Butucaru's compilers produce.

1.9 Executing the Generated Code

The code generated from an Esterel synchronous program is generally executed in a standard sequential framework where events come asynchronously. Since its execution does take time, running the code in a way that respects the synchronous requires some care. It basically requires the definition and implementation of an external *execution shell* that gather inputs from the

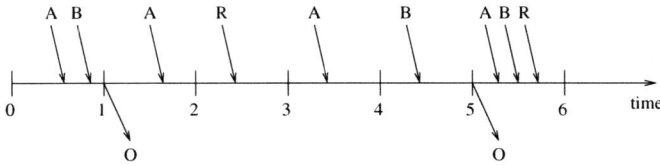

Figure 1.7: A possible run of ABRO in an asynchronous environment

environment, builds logical input events for the reaction functions, and performs the actual output actions. To date, all Esterel code generators have relied on a simple execution shell structure that has remained practically unchanged since its initial definition for Esterel V2.

The generated code is organized around a *reaction function* that effectively computes the transition function of the underlying FSM. The reaction function is called at each logical instant. Conceptually, it takes environmental inputs and the current state as parameters and produces outputs and the next state. The actual implementation is a bit more detailed; see Appendix C.

The execution shell is responsible for repeatedly triggering the reaction function. There are a variety of choices for how to do this. The reaction function may be called periodically, typical in automatic control; when any input signal arrives, an event-driven style; when a particular signal arrives, in which case that signal acts as a clock; or immediately after the previous reaction has been computed. The only restriction imposed by the synchrony hypothesis is that the computations of successive reactions do not overlap and that the state of the inputs does not appear to change during a reaction. Checking that reactions do not overlap in the periodic case can be done done through worst-case execution time (WCET) analysis.

One way to keep the inputs constant is for the execution shell to buffer inputs, effectively synchronizing them. One simple strategy is to freeze the input buffers when the reaction function is called and start recording input events again when it returns. If the implementation respects the synchrony hypothesis, these may be simple single-place buffers that simply record whether an event has been observed since the last instant. Figure 1.7 shows a possible run of a software implementation of ABRO that corresponds to the synchronous trace of Figure 1.2. The reaction function is called at the numbered times. Of course, other execution schemes are possible.

1.9.1 Existing Solutions

The execution shell assumed by Esterel compilers has remained practically unchanged from the first compilers. Its API, called the *host language interface* is a *de facto* standard shared by the different existing compilers. We present it in Appendix C.

The main research effort in Esterel compilation has focused on the synthesis of reaction functions. The goal is a practical compilation process that generates correct, fast, and compact reaction functions for all Esterel programs. In practice, different compilation schemes have different advantages and disadvantages.

More recent techniques attempt to combine the advantages of previous ones. They generate code that simulate the reactive features of the language at run-time. We shall call such a target code *control-flow code* of the Esterel description. The control-flow code is small—comparable in size to the circuit—and evaluates only active statements in an instant—as the FSM code does. The difficulty comes in scheduling program fragments that respect the language semantics, more specifically the causality relations. Code to ensure the correct activation of such fragments is a significant part of the generated code.

As the example of Figure 1.6 shows, efficient sequential code is obtained by *statically scheduling* the reaction operations. The results are excellent: the code is smaller than that from the circuit technique and approaches the speed of explicit FSM code. The drawback is that this technique cannot handle all correct Esterel programs. Specifically, to statically schedule an Esterel program, the control- and data-flow structure of the program must be statically acyclic, a stronger restriction than necessary. In particular, the order may depend on the program state and inputs, meaning completely static scheduling is impossible. More importantly, each translation technique defines "acyclicity" of the control- and data-flow structure in a slightly different way and optimizations may add or remove cycles. Thus, the class of accepted programs depends on the chosen compiler and optimization level. We shall give a detailed overview of existing control-flow code generators in Chapter 7.

2
The Esterel Language

Esterel is an imperative synchronous language for the programming of reactive applications. Developed by a team lead by Gérard Berry, its first version was defined more than twenty years ago [12]. This book describes version 5 (V5) of the language [7, 13]. The current commercial version of Esterel is 7 [29]—a broad syntactic extension of V5 with little change to its semantics.

This chapter is devoted to an intuitive presentation of the language that omits details and certain constructs. We cover these later in the reference manual in Appendix B. Furthermore, we defer a detailed discussion of the Esterel semantics to the second part of the book. Instead, we concentrate on the description of the mechanisms that allow the definition of high-level reactive behaviors.

The presentation is roughly divided in two: The first part focuses on the expressive power of the language; the second covers practical issues such as modular development and interfacing with the real world. The first part starts with an overview of the language constructs, followed by the definition of the *primitive* statements of Esterel. Next, we show examples to familiarize the reader with the notions of *cyclicity*, *schizophrenia*, and *reincarnation*, which present key challenges in the code generation process.

The second part of the chapter defines the notions of *host language* and *module* and defines the general structure of an Esterel specification.

2.1 Syntax and Naïve Semantic Principles

An Esterel program consists of a declaration header followed by a reactive body statement. The header defines the *interface* of the program: input and output signals, data types, variables, and external data manipulation routines. Esterel offers a wide variety of statements that may appear in the imperative body. Among them are the primitive language constructs, also called *kernel statements*, on which the language semantics is defined. The rest of the statements are syntactic sugar derived from the kernel statements, which prove convenient for the development of practical applications.

We now take a short tour of the language features. We have divided our presentation into several steps, each introducing a certain class of language constructs. The complex issues are clarified at each step through small examples.

Control flow

Esterel provides classical control-flow constructs. It has a no-operation statement (**nothing**), parallel ($p \parallel q$) and sequential ($p\ ;\ q$) composition, and iterative sequential loop (**loop** p **end**). Once executed, a parallel statement terminates when all of its branches have terminated. A loop runs forever unless terminated by an explicit exit or preempted by its environment. The body of a loop is not allowed to terminate instantly when first executed to prevent unbounded computation and guarantee the completion of synchronous reactions. Compilers reject programs that contain such loops.

The trap statement "**trap** T **in** p **end**" defines an instantaneous exit point for its body. The termination of the trap statement is triggered by "**exit** T" statements, where T is a trap label. If "**exit** T" executes inside p, the entire trap statement terminates immediately. Notice that we use the specific "trap" word instead of the better known "exception" word used in some languages. We think that "exception" is much less precise, since it also covers arithmetic exceptions such as zero-divide that are not covered by trap.

Classical sequential variables can be defined in Esterelas follows.

    var v := *expr* : *type* **in** p **end**

Such variables can be assigned

    v := *expr*

and tested

    if *cond* **then** p **else** q **end**

Here, v is a variable name, *expr* is an expression, and *cond* is a Boolean expression involving variables, such as "v < 10." The variable declaration statement defines the sequential variable v of type *type*. The scope of v is the statement p. When the declaration statement is executed, the initializing expression is computed and the result is assigned to the variable as initial value before control is passed to the body p.

Variable initialization is required to ensure the macrostep determinism of the language, but it can be skipped to simplify the code when no initialization is needed (i.e., when an assignment is always executed before the variable is read).

Concurrent access to sequential variables is restricted syntactically. When one branch of a parallel statement assigns a sequential variable, none of the other branches is allowed to read or assign the same variable. Esterel thus prohibits both write-write and read-write conflicts. For example, Esterel compilers reject the following compound statements.

```
w := 10 || w := 11

v := 10 || w := v

if u = 0 then v := 10 end || if u <> 0 then w := v end
```

The first example contains a write-write conflict. A read-write race is present in the second. The third example emphasizes the syntactic nature of the restriction enforced by compilers. While the guards prevent the read and write actions from occurring simultaneously, the fragment is still rejected.

Division of behavior in reactions

The **pause** statement and its numerous derivatives are used to divide the behavior of the program into successive instants. When the sequence

```
emit A ; pause ; emit B
```

is started, it emits A, then *pauses* (freezes) until the next execution instant, where it *resumes* execution to emit B and terminate.

Signal communication

Parallel branches may communicate (synchronize with each other) through signals. Two types of signals exist in Esterel:

- *pure signals*, which only have a *status* of *present* (*true*,1) or *absent* (*false*,0); and

- *valued signals*, which carry a typed *signal value* in addition to the Boolean status.

The statement "**signal S in** p **end**" declares the pure signal S for the scope of p. It is emitted by "**emit S**" and its status can be tested with a wide variety of constructs, including the following.

- "**present** *signal-expr* **then** p **else** q **end**" is the regular branching test, where *signal-expr* is a *signal expression*, obtained by combining signal names with the Boolean operators **and**, **or**, and **not**.

- "**suspend** p **when** *signal-expr*" is the primitive preemption test. In instants where the test expression is true, the **suspend** statement prevents its body from running. The state of the body is held unchanged in such instants.

- "await *signal-expr*" is a derived statement that blocks execution until the signal expression is true. It always blocks for at least one instant.

- "abort *p* when *signal-expr*" is a derived preemption statement. When the test expression is true, the abort statement preempts its body, resets its state to inactive, and gives control in sequence.

When the statement declaring the pure signal S is started or resumed (e.g., in a loop), the status of S is unknown. It becomes *present* for the current instant as soon as S has been emitted. It is set to *absent* only when it can be determined without speculation that none of the "emit S" statements can be executed in the current instant. A test statement can read the status of S only after its presence/absence has been established.

The following example shows how the simple signal- and data-handling statements introduced earlier can be used to define a signal occurrence counter. The Esterel statement await 10 I has the following expansion over these simpler constructions.

```
trap T in
  var v := 0 : integer in
    suspend
      loop
        pause ;
        v := v + 1 ;
        if v = 10 then exit T end
      end
    when [not I]
  end
end
```

When this statement starts, the integer variable v is set to 0 and control freezes at the pause statement. In all subsequent instants, the signal I is either absent, which suspends the inner loop, or present, in which case v is incremented and the loop body is restarted. When v equals 10, the trap T is exited, causing the whole statement to terminate instantly.

Valued signals

A valued signal S of type *type* can be seen as a pair consisting of a pure signal with the same name (S) and a persistent shared data variable, named ?S, of type *type*. The valued signal S is declared with a construct of the form

signal S := *expr* : combine *type* with *c* in *p* end

The data expression *expr* of type *type* gives the initial value of ?S. We will explain later how an associative and commutative *combine function*

$$c : type \times type \to type$$

is used to deterministically handle write-write conflicts for valued signals.

Note that Esterel compilers do not require the initialization of valued signals, nor the use of a combine function. This allows the user to write simpler code in cases where it is clear that the initial value is never read or when a valued signal cannot be emitted twice during a reaction. In the examples in this book, we shall use the abbreviated actual syntax where convenient. However, all the semantic developments use the full version of the statement with initialization and a combine function.

Pure signal emission and shared variable assignment can only be performed together by the *valued signal emission* statement "emit S(*expr*)." The type of the data expression *expr* must match that of the signal S.

The pure signal component may be used in signal test expressions as if it were a non-valued signal, and the shared variable ?S may be used in data expressions. Unlike sequential variables, valued signals allow data communication between parallel branches. We shall see later that they are subject to synchronization rules that may lead to deadlocks.

While the pure signal and the shared variable are syntactically linked for programming style reasons, the situation is different at the semantic level. In fact, the two objects follow different synchronization rules and we will dissociate them completely for analysis and code generation. The pure signal part follows the lazy synchronization rules defined on page 18. By contrast, the shared variable is subject to strong synchronization to ensure that its value is the same in all the expressions that read it throughout each instant. The variable can be read in an instant only after *all* corresponding "emit" statements have either been executed or ruled out as unreachable. In instants where the valued signal is not emitted, the value of the variable does not change. If exactly one emission of the signal occurs during an instant, then the value is given by the evaluation of the expression of the single emit statement that is executed. When multiple emissions occur, the corresponding values are combined into the new value of ?S using the associative and commutative combine function *c*.

In the example below, the value of the variable v will be 11 in the first execution instant, 11 in the second, and 7 in the third.

```
signal S : combine integer with + in
  emit S(5) ; pause ; pause ; emit S(7)
||
  emit S(6) ; pause ; pause
||
  loop v := ?S ; pause end
end
```

The values 5 and 6, emitted for S in the first instant, are combined with the function "+" before the execution of the assignment.

The next example is incorrect because it contains a *causality cycle* that amounts to a deadlock: the value of S can only be read after its emission while the sequence (;) requires the assignment to be executed first.

```
v := ?S ;
emit S(6)
```

The following artificial example shows how valued signals and sequential variables are used together to compute complex functions, here, the suite of partial sums of the series $\Sigma_n(2 * I_n + I_n^2)$, where the values $\{I_n\}_n$ are given to the program through the integer input signal I:

```
module SUM:
input I : integer ; output SUM : integer ;
signal TERM : combine integer with + in
  suspend
    loop pause ; emit TERM(2*?I) end
  ||
    loop pause ; emit TERM(?I*?I) end
  ||
    loop
      var acc :=0 : integer in
        pause ;
        acc = acc + ?TERM ;
        emit SUM(acc)
      end
    end
  when [not I]
end module
```

The program initializes itself in the first instant: the status of I is ignored, and 0 is not emitted. In subsequent instants where I is not emitted by the environment, the program does not change its state. When I is present, the accumulator acc is updated and the computed sum is passed back to the environment through SUM. The value of I is read by the first two parallel branches, which compute each one part of the term. The results are then combined and read by the third branch, which updates the partial sum and sends it to the environment.

2.2 The Kernel Esterel Language

Kernel Esterel is the subset of the Esterel language defined by the primitive (kernel) Esterel statements. We use it to simplify the formal definition of the language semantics; only the semantics of the kernel statements are given directly. The semantics of the remaining derived statements are defined in terms of kernel statements. For instance, the derived statement "await S" is expanded into

```
trap T in
  loop
    pause;
    present S then exit T else nothing end
  end
end
```

The semantics of an Esterel program is given by its expansion into a Kernel Esterel program.

The Esterel translation schemes defined in this book are also based on the kernel language, as structural translation patterns are only defined for the language primitives. The various compilers actually translate a slightly larger number of constructs to generate better code. For instance, the translation of "await S" is usually based on the following (non-kernel) expansion:

```
suspend
  pause
when [not S]
```

Such implementation details, covering non-kernel constructs, will be presented in Appendix A. There, we shall cover signal expressions such as "not S," the pre operator, etc.

The kernel-based approach is natural to Esterel, as many of the language features are "syntactic sugar" meant only to facilitate the task of the developer. The same approach is used by Berry [7] to define the semantics and the hardware translation of the data-free subset of Esterel, henceforth named Pure Esterel.

To obtain a full set of primitives, we extend Berry's Pure Esterel Kernel with data-handling statements. Of the sixteen primitives that form Kernel Esterel, eleven are the Pure Esterel primitives listed in Table 2.1.

The choice of data manipulation primitives is not obvious, as the valued signals of Esterel are difficult to model. We mentioned before that the valued signals are hybrid data/control constructs that follow two synchronization rules: lazy for the status and strict for the value.

The Pure Esterel signal and the shared variable that compose a valued signal are syntactically linked in declaration and emission, but are otherwise independent. The signal is present and ready to be read as soon as it has been emitted. The Boolean operations over signal statuses are lazy: A or B is true as soon as one of A and B has been emitted. However, the shared variable can be read only after *all* the emissions of the valued signal have been either executed or ruled out by the control flow. The operations over signal values are strict: if A and B are Boolean valued signals, then ?A or ?B can be computed only when both ?A and ?B are known.

The difference is illustrated by the following two short and very artificial examples. The first one is semantically correct:

nothing	signal S in p end
pause	emit S
p ; q	present S then p else q end
$p \parallel q$	suspend p when S
loop p end	trap T in p end
	exit T

Table 2.1: The primitives of Pure Esterel, as defined by Berry [7]. Here, p and q are Esterel statements, S is a signal name, and T is a trap name. We will extend the trap notation in Section 3.3.4 to facilitate the presentation of semantic and code generation issues.

```
signal A : combine boolean with or in
  emit A(true) ;
  present A then emit O end ;
  emit A(true)
end
```

When the control reaches the signal test, we already know that the A is present. Thus, we emit O and, a second time, A. The second example is incorrect:

```
signal A : combine boolean with or in
  emit A(true) ;
  if ?A then emit O end ;
  emit A(true)
end
```

Here, the test of ?A occurs while the second emit statement has not yet been executed. The causality cycle cannot be resolved in this case and therefore the compiler rejects this example.

In the definition of the kernel language, we separate the pure signal part of a valued signal from the data part. We do this by introducing independent primitives for the manipulation of shared variables. These primitives are not part of the Esterel language per se; they are only used for semantic analysis purposes and for defining the code generation schemes.

Here, we list the five data-handling kernel statements. In these definitions, v and v_i are non-shared variable names, s and s_i are shared variable names, and p and q are Esterel statements.

- "var v := $expr$: $type$ in
  p end" is the non-shared variable definition, defined earlier in this chapter.

- "shared s := $expr$: combine $type$ with c in p end" corresponds to the data part of an Esterel valued signal declaration and is not actual Esterel syntax. It declares p as the scope of the persistent shared Boolean variable s. The variable has the type $type$ and is initialized with the expression $expr$ when the statement is started. Write-write conflicts are resolved by composing the concurrent write actions (defined below) using the associative and commutative binary operator c—the *combine function* of s. Read-write conflicts are resolved by requiring that all write actions are either executed or ruled out by the control flow before s is read.

- "$v := f(s_1, \ldots, s_n, v_1, \ldots, v_m)$" awaits the completion of the computation of its shared variable arguments $(s_1, \ldots, s_n)$, and then assigns the value of the function to the sequential variable v. Recall that the synchronous abstraction assumes that the computation of f takes zero time, which means that it fits inside the computation of the reaction.

- "$s <= f(s_1, \ldots, s_n, v_1, \ldots, v_m)$" awaits the completion of the computation of its shared variable arguments, and then generates a write action on the shared variable s. Again, the computation of f is assumed to be performed in zero time. This is not actual Esterel syntax.

- "if v then p else q end" is the classical conditional with the Boolean non-shared variable v as condition.

All the constructions of Esterel can be represented through macro-definitions over the kernel language. Appendix B gives the full set of expansion rules. Here, we only show how valued signals are expanded into combinations of pure signals and shared variables.

Each valued signal S is represented using a pure signal with the same name (S) and a shared variable named Svar. Then, the declaration

  signal S := $expr$: combine $type$ with c in p end

is expanded into

  signal S in
    shared Svar := $expr$: combine $type$ with c in
      p'
    end
  end

where p' is obtained by replacing all the instances of ?S in p with Svar and all the instances of "emit S($data\text{-}expr$)" with

  emit S ; Svar <= $data\text{-}expr$

24   THE ESTEREL LANGUAGE

Notation

For the rest of this book we shall abbreviate the variable declaration statements as follows.

- "**var** v **in** p **end**" declares the non-shared variable v of type $type_v$ and initialing expression $init_v$.

- "**shared** s **in** p **end**" declares the shared variable s of type $type_s$, initializing expression $init_s$, and combine function c_s.

We will use these abbreviations in the semantic formulas, but not in the Esterel code examples. Note that the **var** construct is standard Esterel but the **shared** construct is an internal concept.

2.3   Esterel Through Examples

Here, we present through examples the main semantic problems that will require attention in the later parts of the book. Later, we will analyze some of the examples extensively when considering certain code generation techniques. Each example we present here is complete in the sense that it includes the declaration header. We will explain the behavior of each example intuitively; later, we will be more formal.

Esterel is a parallel language. The easy-to-use signal communication supports a programming style based on fine-grained parallelism and synchronization, where complex behaviors are constructed hierarchically. Our first example is a Pure Esterel program designed to include many of its these features in little space.

```
module MainExample:
input I, J, KILL, SUSP; output O; % interface declarations
suspend
  trap T in % performs the preemption
    signal END in
      loop % basic computation loop
        await I ; emit O ; await J ; emit END
      end
    ||
      % preemption protocol, triggered by KILL
      await KILL ; await END ; exit T
    end
  end;
when SUSP % suspend signal
end module
```

The program models a cyclic computation (like a communication protocol) that can be interrupted between cycles and suspended. When started, the

Instant	Inputs	Outputs	Comment
0	any		All inputs discarded
1	I	O	
2	KILL		Preemption protocol triggered
3			Nothing happens
4	J,SUSP		Suspend, J discarded
5	J		END emitted, T exited, program terminates

Figure 2.1: A possible execution trace of MainExample

await statement waits for the next clock cycle where its signal is present, i.e., it always delays at least one cycle. All the other statements in our example execute in a single clock cycle, so the await statements are the only places where control can be suspended between instants (they preserve the *state* of the program between cycles). One consequence is that the I and J signals must arrive in different clock cycles, otherwise J will be ignored.

The loop is preempted by the trap statement trap when "exit T" executes. In this case, the trap terminates instantly, control is passed to the next instruction in sequence, and the program terminates. The preemption protocol is triggered by the input signal KILL, but T is exited only when END is emitted. The program is suspended—no computation is performed and the state is kept unchanged—in cycles where the SUSP signal is received. A possible execution trace for this program is given in Figure 2.1.

The computation of each reaction is instantaneous but *causal*. The elementary computations defining the reactive behavior of a program are connected through three types of causal dependencies: control flow (sequencing), signal communication, and access to shared variables. In MainExample,

- sequencing requires that the statement "exit T" is always executed *after* the preemption test on the signal END; and

- signal synchronization through END requires that the test on END is executed after the test on J, whenever the two are executed in the same instant.

The computation of every instant is performed causally, but not all causal dependencies apply in every instant. For instance, the dependency on END only applies when "await J" and "await END" are both active.

Understanding the system of causal dependencies is essential to generate sequential code. When the dependencies are acyclic, the elementary computations of the program can be totally ordered in a way that satisfies the causal dependencies in *all* instants. Thus, good sequential code can be generated using *static scheduling* techniques. Like most of our examples, MainExample falls in this category.

Cyclicity

In some cases, however, the causal dependency system is cyclic. We say that the program is *cyclic*, or that it contains a *causality cycle*. The following example illustrates this.

```
signal S,T in
  present I then emit S else emit T end
||
  present S then call f1()() ; emit T end
||
  present T then call f2()() ; emit S end
end
```

Here, the status of signal S *statically* depends on the status of T and vice versa. Consequently, the order in which the functions f1 and f2 are called depends on the input signal I. The function calls must be dynamically scheduled.

Cyclic programs are difficult to handle. Proving their correctness requires expensive analysis. Perhaps more importantly, it is difficult to generate efficient code for them. In the general case, dynamic scheduling is required and, as in the following example, not-yet-executed code may have to be evaluated *before* execution to determine the absence of a signal.

```
signal B1, B2 in
  emit A ;
  present B2 then emit O1 else emit O2 end ;
  present A then emit B1 else emit B2 end
end
```

Here, after the emission of A, the execution is blocked on the test on B2. But A is present, so the test on it cannot take the "else" branch, so B2 is not emitted, so it is absent. Then, we execute the test on B2 and emit O2, we pass in sequence, we execute the test on A and emit B1. Note that determining that B2 is absent involves no speculative computation and, since the control is blocked, no actual execution of code. Instead, computation proceeds by invalidating test branches based on the status of signals that have already been computed. We call this kind of forward evaluation *potential computation*.

But the first problem related to cyclic programs is the formal definition of what "cyclic" means. We shall see later that the definition is generally given at the level of graph-based intermediate compiler representations. Unfortunately, this means slightly different definitions are associated with each intermediate representation and therefore each variant compilation technique. The following program, for instance, is considered acyclic by some compilers and can be statically scheduled, while other compilers reject it as cyclic.

```
module Monster:
output O;
signal S,A,B in
  emit S;
  [
    present A else pause end
  ||
    present B else emit O end
  ];
  present S else emit A end;
  emit B
end
end module
```

The exact reasons for this behavior are complex and their understanding requires knowledge of translation details that we will present later. In Section 9.1, we analyze this problem and derive a partial answer to what cyclic really means.

Cyclic Programs Can Be Useful Our last correct cyclic program is the bus arbiter example. This handles the bus access requests of n users. At each instant, control is given to exactly one of the users that request it. It is a fair arbiter: a user requesting control during n consecutive instants receives control at least once. For simplicity, we have written the code for $n = 2$, but it can easily be modified to support any fixed number of users.

```
module Cell:
input REQ; output ACK;
input GRANT_IN; output GRANT_OUT;
input TOK_IN; output TOK_OUT;
  loop
    present TOK_IN then pause;emit TOK_OUT
    else pause end
  end
||
  loop
    present [TOK_OUT or GRANT_IN] then
      present REQ then emit ACK else emit GRANT_OUT end
    end ;
    pause
  end
end module

module Arbiter2:
input REQ1,REQ2;
```

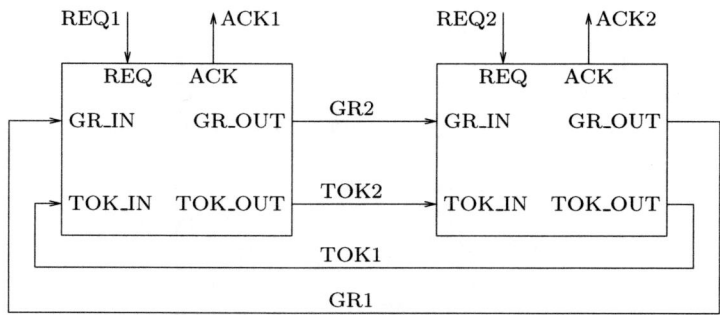

Figure 2.2: Interconnections between `Cell` modules in `Arbiter2`

```
output ACK1,ACK2;
signal GR1,GR2,TOK1,TOK2 in
  emit GR1 ; emit TOK2
||
  run Cell[signal REQ1/REQ,ACK1/ACK,GR1/GRANT_IN,
          GR2/GRANT_OUT,TOK1/TOK_IN,TOK2/TOK_OUT]
||
  run Cell[signal REQ2/REQ,ACK2/ACK,GR2/GRANT_IN,
          GR1/GRANT_OUT,TOK2/TOK_IN,TOK1/TOK_OUT]
end
end module
```

Each user communicates through `REQ` and `ACK` signals with one of the instances of the `Cell` module. A token is used to ensure fairness. In each instant, the token is passed from one module to the other through the signals `TOK_IN` and `TOK_OUT`. During an instant, a request to a `Cell` having the token determines an acknowledgment. However, if access is not requested at that cell, another cell may grant access upon request. The cells grant each other control through the signals `GRANT_IN` and `GRANT_OUT`. Figure 2.2 shows how the two `Cell` modules are connected. The first parallel branch of the main module initializes the system. Here, the static causality cycle is the dependencies that link the signals `GR1` and `GR2` by means of causal dependencies internal to the `Cell` modules.

Causality errors

All previous cyclic examples were correct—at each instant the static cycle is broken. This is not the case for all programs, however. As in many concurrent formalisms, it is easy to build Esterel programs that deadlock. Such programs are considered incorrect and are usually called *non-constructive* to reflect that they cannot be evaluated completely in a constructive manner. The exact definition of non-constructivity shall be given in the second part of this book.

Schizophrenia and Reincarnation

The last semantic property that we describe here is the *reincarnation* of signals and control. The following example illustrates signal reincarnation.

```
loop
  signal S in
    present S then emit O1 else emit O2 end
    pause ;
    emit S ;
    present S then emit O1 else emit O2 end
  end
end
```

In the first instant, O2 is emitted and the **pause** statement activated. When the statement is resumed in the second instant, the signal S is emitted, and the second test on it emits O1. Then, the **signal** statement terminates and is instantly restarted with a fresh copy of S. The fresh signal is not emitted so it is absent. The program then emits O2. We say that the signal S has been reincarnated because two independent copies of it exist within an instant. We also say that it is schizophrenic because it has more than one status in an instant.

More complicated examples of reincarnation are possible. For example, a single statement can be executed with different inputs and with different results several times in an instant. In the following example, the presence test on S is executed twice in the second instant, with different results.

```
module REINCARNATION:
output O1,O2;
loop
  signal S in trap T in
    pause ; emit S ; exit T
  ||
    loop
      present S then emit O1 else emit O2 end ;
      pause
    end
  end end
end
end module
```

We say that the statement is schizophrenic and that it is reincarnated once in the execution of the second instant.

In the previous examples, only two instances of a signal or statement coexist inside an instant. But Esterel programs can be built that require any bounded number of instances of a signal or statement. Consider the following example, inspired from Berry [7].

```
module MULTIPLE_REINCARNATION:
output O1,O2,O3,O4 ;
loop
  trap T1 in signal S1 in
    pause ; emit S1 ; exit T1
  ||
    loop
      trap T2 in signal S2 in
        pause ; emit S2 ; exit T2
      ||
        loop
          present S1 then
            present S2 then emit O1 else emit O2 end
          else
            present S2 then emit O3 else emit O4 end
          end ;
          pause
        end
      end end
    end
  end end
end
end module
```

In the second execution instant it will execute the "**present S1**" statement 3 times, in different signal contexts, resulting in the emission of the signals O1, O2, and O4.

This form of duplication might appear not to comply with the synchronous model, which requires that every signal has exactly one status during a given execution instant. However, the reincarnation mechanisms included in the semantics of Esterel allows the correct interpretation of such compact definitions, which allow for a bounded number of instances of a signal or statement to be evaluated inside an instant. Reincarnation can sometimes simplify the task of the programmer by allowing more compact representations of algorithms. In general, various instances (incarnations) of a schizophrenic signal or control operator must be expanded through a bounded unrolling before any serious analysis or code generation can take place.

While the execution of different incarnations of a statement is always realized in a given order, reincarnation is not a simple signal re-initialization and loop: The potential function computation may evaluate future incarnations of a statement/signal before control enters them. For any natural number n, examples can be constructed where n future incarnations of a statement must be explored in order to make a decision about the currently executing one.

2.4 Host Language

Esterel relies on general-purpose languages such as C or C++ in two ways. First, for portability, most Esterel compilers generate C instead of assembly or some other executable representation. Such generated code provides an interface for passing events between the environment and the running program. Second, Esterel allows the use of data types and functions defined externally (e.g., in C) to be used within an Esterel program.

The language used for these purposes is called *host language*. The *host language interface* is the application programming interface (API) that specifies how data types and routines defined in the host language are imported for use in Esterel, and how the host language code can trigger synchronous reactions, emit and receive signals, etc.

To allow interaction between the Esterel code and the host language, the host language interface involves both Esterel constructs (the program interface, defined in the next section), and host language constructs. The C language API is presented in Appendix C.

2.5 Program Structure and Interface

The basic programming unit in Esterel is the *module*. All declarations and statements must be part of a module, so there are no global declarations of types, variables, signals, etc. Nested module declarations are also not allowed. An Esterel specification is a set of modules contained in one or more files. The modules of a specification are organized hierarchically. Using the **run** pseudo-statement, described in this section, a unique *root module* can instantiate other modules, which can in turn instantiate other modules. In this respect, modules behave like parameterized templates. Recursive module instantiation (i.e., a **run** statement directly or indirectly calling its enclosing module) is prohibited.

The general structure of a module is

> **module** *module-name* :
>   *interface*
>   *body*
> **end module**

Program body

The *body* of a program is an Esterel statement that defines the behavior of the module. To facilitate the hierarchical composition of modules, body statements may not have free traps, that is, **exit** statements that have no corresponding trap declaration super-statement; or free sequential variables.

Thus, inter-module communication can be performed only through signals. The following statement cannot be a program body since it has a free trap and two free sequential variables.

```
v := 10 ;
w := v+1 ;
exit T
```

Program interface

In this section, we describe the interface, which declares two types of objects:

- *input and output signals*, which are the only communication lines between the module and its environment, as well as *signal relations* specifying simple assumptions over the behavior of the environment; and

- the data structures and data handling routines that are defined and implemented in the host language and used in the module—*types, constants, procedures, functions,* and *tasks*.

Declarations can be freely mixed, but an object must be declared before it is used.

2.5.1 Data Handling

The type system of Esterel is rudimentary, with no type constructors or subtyping. A type declaration simply specifies a name. Type checking is strict, based on the name. The following example declares three types—T1, T2, and T3.

```
module Example:
  type T1, T2;
  constant C1 : T1, C2 = false : boolean;
  type T3;
  procedure P(boolean,T3)(T2), PrintT1()(T1);
  function Length(string):integer;
  task Tsk(T2,string)(string);
  ...
end module
```

All such declared objects, including types, must be defined in the host language, following the rules of Appendix C.

Constant declarations specify the constant name, its type, and possibly a value, which is another constant or a literal of one of the predefined types.

Procedure declarations specify the procedure name and two formal parameter lists. The first contains parameters that are passed by reference; the second contains pass-by-value parameters.

Function declarations specify the function name, its parameter list, and its return type. All function parameters are passed by value.

Each execution of a function, procedure, or predefined operator is assumed to execute within a single clock cycle. Thus, it can be considered "instantaneous," is non-interruptible, and can be fully described in the Esterel semantic framework. However, it is often the case that complex data computations or system/environment interactions do not fit within a single execution instant. The Esterel language offers specific constructs, called *tasks*, adapted to these cases. As the previous example showed, tasks are declared much like procedures, the only difference being the use of the **task** keyword instead of **procedure**. Details concerning task semantics and their use are in Appendix A.4.

Predefined types and operators

Esterel defines common types, constants, and operators. There are five predefined types: **boolean**, **integer**, **float**, **double**, and **string**.

The predefined Boolean constants (literals) are **true** and **false**.

Details of the predefined numerical types depend on the host language implementation (e.g., C). In the INRIA Esterel V5 compiler, legal numeric literals are specified by the regular expression

$$[0\text{--}9]^+ \ ([.][0\text{--}9]^*)? \ ([e|E][-|+]?[0\text{--}9]^+)?$$

Integer literals are specified by the regular expression $[0\text{--}9]^+$, but correct code is generated only for literals that fit the target architecture. Floating point literals contain one decimal point and/or an exponent. To differentiate **float** literals from **double** literals, an "f" is appended to the former. Thus, 0.1 is a **double** literal, whereas 1e3f is a **float** literal. Current implementations of Esterel do not check the size of floating-point literals; they simply copy them into the host language.

Strings are single-line sequences of characters delimited by double quotes. The only escape sequence is "", which is transformed into a single double quote. Different implementations introduce various limitations, such as the maximum length of a string. See Appendix C for details.

For all types, including user-defined ones, the assignment :=, equality =, and inequality <> operators can be used. Implementations of these operators for user-defined types must be provided in the host language.

Boolean constants and variables can be manipulated using **and**, **not**, and **or** operators. As usual, **not** binds tighter than **and**, which binds tighter than **or**.

The built-in numeric types support the +, -, *, /, <, <=, >, and >= operators. The modulo operator **mod** can be used only on integers. Unary - has the highest priority, followed by *, /, and **mod**, followed by + and binary -. The semantics of each operator is the same as the corresponding one in the host language. There is no implicit conversion among numerical types.

2.5.2 Signal and Signal Relation Declarations

Every Esterel module has one predefined signal called **tick** that is present in every cycle. It can be thought of as the clock of the module.

Every signal that is used must either be declared in the module body or in its interface. Interface signals represent the only means for an Esterel module to communicate with its environment. Five classes of interface signals exist, distinguished by the keywords **input**, **output**, **inputoutput**, **return**, and **sensor**. Here, we give the intuition behind them; we defer the formal definition to Appendix B.

The most common are **input**, **output**, and **inputoutput** signals, which can be read and emitted within the module body. In addition, **input** signals can be emitted by the environment, **output** signals emitted within the module body are visible to the environment, and signals of class **inputoutput** behave as both **input** and **output**. We will define the module-environment communication mechanisms in the next section, where we describe the the **run** pseudo-statement.

The signals of classes **return** and **sensor** can be read but not emitted by the module body. Signals of class **return** are used to mark the completion of *tasks*, described in Appendix A.4. Sensor signals represent pure data input. Their present/absent status cannot be read, only their value. In this sense, they can be seen as shared variables that cannot be emitted by the module.

A signal declaration specifies its name and class. A declaration of a valued signal also specifies its type, the initial value, and its combine function. The initial value and combine function can be omitted.

Signal relations declare simple relationships among the present/absent status of **input**, **inputoutput**, and **return** signals. It is assumed that the environment guarantees these relationships, making it possible to avoid specifying irrelevant behaviors. Two types of properties can be specified: exclusion and implication. For example,

```
relation I => J, I # K # L ;
```

states J must be present when I is present, and that I, K, and L are mutually exclusive, i.e., if one is present, the others are not.

2.5.3 The *run* Pseudo-Statement

Due largely to its complex semantics, the Esterel language has no mechanism for separate compilation or pre-compiled component libraries. However, users may assemble modules to build complex specifications through *submodule instantiation*. The **run** pseudo-statement instantiates such modules. As suggested by the term *instantiation*, the mechanism is similar to macro or template instantiation. Here, the templates are the modules themselves, and the template parameters are the signals, etc. declared in the module interface. Instantiating a module inside another consists of expanding the **run**

```
module FIFO:
 type FifoType ;
 input   DataIn  : FifoType ; output BlockIn   ;
 output DataOut : FifoType ; input   BlockOut ;
 signal Data1:FifoType,Data2:FifoType,Block1,Block2 in
   run Cell[type FifoType/CellType ;
           signal Data1/DataOut, Block1/BlockOut]
 ||
   run Cell[type FifoType/CellType ;
           signal Data1/DataIn,  Block1/BlockIn,
                  Data2/DataOut, Block2/BlockOut]
 ||
   run Cell[type FifoType/CellType ;
           signal Data2/DataIn,  Block2/BlockIn ]
 end signal
end module

module Cell:
 type CellType ;
 input   DataIn  : CellType ; output BlockIn   ;
 output DataOut : CellType ; input   BlockOut ;
 loop
   var v : CellType in
     await immediate DataIn ; v:=?DataIn ;
     abort pause ; sustain BlockIn when [not BlockOut] ;
     emit DataOut(v)
   end var
 end loop
end module
```

Figure 2.3: A 3-cell FIFO in Esterel. It is required that the environment respects the FIFO protocol by not presenting new input on DataIn when BlockIn is present (such input would be lost).

statement of the *caller module* into the statement body of the *instantiated module*. The body of the instantiated module is rewritten in the process to take into account the actual parameters of the run construct.

Here, we present the intuition behind the instantiation mechanism; we defer the formal semantics to Appendix B.

A run statement consists of the run keyword* followed by the module name and an optional renaming list. The renaming list is delimited by square brackets and its items are separated by semicolons. Each item starts with one of the keywords type, constant, function, procedure, task, or signal, and is followed by a list of comma-separated *renamings* of the correct type.

Figure 2.3 illustrates the run construct. This example is a FIFO that deals with messages of type FifoType. The FIFO accepts inputs through DataIn when BlockIn is false and writes output on DataOut when BlockOut is false. It takes one clock cycle for a message to traverse a FIFO cell that is not blocked. This FIFO is of length 3, but it can be easily scaled from 1 to any fixed integer by adding or removing cells—FIFOs of size 1.

Here, the run statement is used to combine three Cell modules into a FIFO. The first instance of run is the entry cell of the FIFO. It sets the type of cell data CellType to FifoType, and links the DataOut and BlockOut signals to local FIFO signals, which are connected with the second cell. Since the DataIn and BlockIn signals of the Cell module are to be linked to identically-named signals of the FIFO module, no explicit renaming is necessary; the connections are made automatically.

Constant, function, procedure, task, and signal instantiations are subject to strict type checks, which are realized after the type instantiations. In the instantiation of constants, either both constants must have identical assigned values (literal), or none may have an assigned value. In the instantiation of valued signals, the type check is performed on the signal type (including the possible combine function). A renaming list can contain as many sections as desired, but no interface object can be renamed more than once.

If an interface object of the instantiated module is not renamed in the run statement, there are two cases: for signals of all types, the caller module must define in its interface a signal with the same name and type; and for data-related objects (types, constants, functions, procedures, and tasks) the caller module can contain no such definition, in which case it is assumed that the environment (another module or the host language) defines it.

Module instantiation

Instantiating a sub-module using the run construct involves two basic operations: the renaming of interface objects according to the parameters of the run statement; and the encapsulation, using *connection code*, of input and output signals according to their direction.

---

*The deprecated copymodule keyword is equivalent.

The first transformation applies to types, constants, functions, procedures, tasks, and to signals of type `sensor`, `return`, and `inputoutput` for which a renaming exists. The renaming takes place hierarchically in the instantiated module body. By default, if a signal in the instantiated module was not renamed, it is connected to an identically-named object in the scope in which it was instantiated. It is an error for no such object to exist.

The second operation applies to *all* interface signals of type `input` or `output`, regardless of whether a renaming is given. For `input` signals, the *connection code* prevents internal emissions from being passed to the exterior. If the input interface signal Loc of type *type* is renamed Glob, the connection code is

```
trap T in signal Loc : type in
  instantiated-body ; exit T
||
  every immediate Glob do emit Loc(?Glob) end
end end
```

For `output` signals, external emissions will not be received in the instantiated module. The connection code is equivalent to

```
trap T in signal Loc : type in
  instantiated-body ; exit T
||
  every immediate Loc do emit Glob(?Loc) end
end end
```

While the previous patterns give the general intuition, the construction of the actual connection code is more complex. In particular, it is assumed that a name assignment step has previously changed the names of the interface signals so that none of them corresponds to the name of a signal used in the `run` renamings*.

In practice, various compilers deviate from these rules. For example, the INRIA Esterel V5 front-end simply ignores signal directions, treating all interface signals as `inputoutput`. The instantiation of a submodule discards the signal relations of its interface.

---

*The patterns would otherwise produce erroneous code if directly applied to statements such as "`run M[signal I/J, J/I, K/K]`".

Part II

Formal Semantics

3
Introduction to Esterel Semantics

The next four chapters, almost a third of the book, are dedicated to a detailed presentation of the semantics of Esterel. This is considerable for a book dedicated to the *compilation* of a programming language, but we believe this is justified by the nature and use of Esterel. First, programs written in the Esterel language have formally defined semantics, which must be respected by any implementation. Second, given the complex control propagation mechanisms of Esterel, only a detailed understanding of the semantics provides the basis for a truly efficient implementation. Indeed, despite its size, our presentation of Esterel's semantics only covers those aspects that are necessary from the perspective of code generation—the constructive semantics and the digital circuit translation. For a broader presentation, see the reference book by Berry [7].

3.1 Intuition and Mathematical Foundations

The semantics of Esterel deal with two intertwined notions: control flow propagation among statements and the propagation of signal statuses. Control flow propagation formally defines which statements are executed in an instant and what they do. Except **pause**, all kernel statements act instantaneously, i.e., execute within a single instant. Signal status propagation determines how signals are set present or absent. The two interact as follows: control flow propagation sets signals present by executing **emit** statements and can also compute signal value; signal statuses and values affect control flow propagation through **present** and **if** statements. In each instant, the status and value of each signal must be determined in a unique way, ensuring behavioral determinism. So we need careful rules about when we may consider a signal absent.

We take the view that a signal S is absent by default and is only present if an "emit S" statement is executed or if it is received from the environment (provided it is an input). The main idea is that we set a signal absent in an instant as soon as the execution status of the program makes the execution of all the instances of "emit S" impossible. Therefore, we do not only propagate the control positively, we also consider which statements *cannot* be executed in an instant to determine when a signal cannot be emitted and set it absent. How to do this has evolved as the Esterel semantics have developed. We describe the current solution, the constructive approach, below. We will mention other solutions later.

Consider the following Esterel statement.

```
signal S, T in
  present I then emit S end
||
  present S then nothing else emit T end
end
```

If the input I is present, S is emitted and set present. The test for S executes its **then** branch and terminates. This tells us that "emit T" will not execute this instant, which means that T can no more be emitted and can be set absent. If I is present, then "emit S" will not run and S can be set absent. The test for S then takes its **else** branch and T is emitted and set present. Of course, the above explanation is still informal. The real difficulty lies in choosing understandable rules for pruning statements that cannot be executed.

Even with consistent rules for pruning unrunnable statements, there may be unsolvable problems called *causality cycles*. Consider the following Esterel statement.

```
signal S in
  present S then emit O1 else emit O2 end ;
  emit S ;
  present S then emit O1 else emit O2 end
end
```

The statement is syntactically correct, and could be given a meaning in a classical control-flow setting. If S were a conventional variable, S would first be found absent by the first **present** statement, emitted and set present by the **emit S** statement, and found present by the second **present** statement. However, this contradicts the synchrony assumption since S should not be considered both present and absent in the same instant. In Esterel this program is considered *causally incorrect*. The constructive approach amounts to prohibiting the emission of a signal from depending on a test for the same signal. Here, what depends means is quite subtle. One can take strong criteria such as dependency graph acyclicity, which is the primary criterion

```
signal R in
  emit S ;
  present R then emit O1 else emit O2 end ;
  present S else emit R end
end
```

Figure 3.1: A correct, but complex Esterel example

for synchronous data-flow languages such as Lustre [38] and has been adopted in Esterel V7*. Alternately, one can employ a more dynamic criteria that only requires input- and state-dependent dependency acyclicity, which is the basis of the *constructive semantics* (see Berry's book [7]). Constructive semantics is elegant, based on the classical mathematical theories of constructive logic rules and three-valued signal propagation, and has deep relations with the safe propagation of currents in cyclic digital circuits, see Berry and Shiple [65]. Another criterion was proposed by Gonthier [5, 35] and used in the Esterel V3 compiler, but is now considered obsolete.

3.1.1 The Constructive Approach

Intuition

As is standard in constructive logics, the constructive semantics of Esterel is based on the propagation of established facts to determine other facts. A fact is knowledge about the presence or absence of signal or about whether a statement will or cannot be executed in the current instant. It is convenient to use the Scott three-valued logic to denote signal facts. In this logic, a signal can be unknown, present, or absent, written $\bot$, 1, and 0. Except for inputs, all signals start off unknown. Constructive propagation can set the status to present or absent. It so happens that once changed, a signal status cannot be changed again in the current reaction, i.e., fact propagation is monotonic. Within a reaction, signal `present` tests effectively block while the status of the tested signal remains undefined. Programs are rejected as non-constructive if all control threads are blocked simultaneously.

Consider the example of Figure 3.1. Execution proceeds as follows: The "`signal`" statement is started and the status of R is set to unknown. Next, signal S is emitted, making it present, and the control reaches the test of R. Since the status of R is undefined, control flow is blocked. However, since S is present, we know the `else` branch of the test of S cannot be executed and may be pruned. Therefore, "`emit R`" cannot be executed, and we can safely conclude R is absent. This unblocks the test of R, which proceeds to emit O2, performs the test of S, and terminates execution.

Applied to the example on page 42, similar reasoning finds the execution

---

*We will say more about this criterion in Chapter 9.

remains blocked forever at the first presence test, since the execution of "emit S" cannot be ruled out by established fact propagation.

Constructive propagation of facts forbids any speculation. Consider the following example.

```
signal S in
  present S then emit S else emit S end
end
```

When this statement is started, the status of S is unknown, and control blocks on the presence test. At this point, the instances of "emit S" cannot be pruned based on known facts, which implies that S cannot be set absent. However, since S is unknown, the control flow is blocked on "present S," implying S cannot be emitted and set present. This last rule is enforced even though every control flow path starting from the blocking state includes "emit S" we are not allowed to speculate on the status of S and deduce that it would always be set present. Therefore, the example is rejected as being non-constructive.*

Fact propagation is a *monotonic* process. Once a fact such as "S is present" has been determined, it cannot change during the rest of the instant. Therefore, complete evaluation of a program is a form of fixed-point computation of facts that runs until no new facts can be determined. At this point, the program is correct w.r.t. a set of inputs if all signals have been assigned a present or absent status.

Theoretical foundations

Berry [7] borrowed the term "constructive" from logic, where *intuitionistic constructive logic* [43, 71] prohibits speculation as a means of reasoning and requires that *all proof objects are effectively built*. Intuitionists reject the law of the excluded middle[†] "$p \vee \neg p = $ true" because it corresponds to speculating over the truth value of p, which may be unknown when the rule is applied.

In an intuitionistic proof, the truth or falsehood of each intermediate proof result is determined in a causal fashion. This form of causality provides a sound semantic basis for automated logical reasoning and for reasoning about computer programs. Consider the following Esterel statement.

```
present I then emit O else emit O end
```

Here, signal O is present (emitted) only after the statement has been started and the status of I has been determined. This is faithfully represented with the following intuitionistic propositional calculus formula.

$$\mathtt{O} \equiv ((\mathrm{START} \wedge \mathtt{I}) \vee (\mathrm{START} \wedge \neg \mathtt{I}))$$

---

*However, this example is considered *logically correct* in the sense of Berry [7]. The logical semantics does not appear practical to compile and is thus only of theoretical interest.

[†]And its applications, such as proof by *reductio ad absurdum*.

Assigning a value *present* or *absent* to O corresponds to proving either O or ¬O, by using the previous formula as a hypothesis and the formulas corresponding to the input valuation, i.e., I for present, and ¬I for absent.

The formula ¬O can be proven only when we assume ¬START, which corresponds to the case where the statement is not executed. The formula O can be proven when we assume START and either I or ¬I, which corresponds to when the statement is executed and the status of I is known. Neither O nor ¬O can be proven with a lesser hypothesis, which would leave the constructive computation blocked.

The constructive approach also gives a clear formal basis for *rejecting programs as semantically incorrect*. In the following statement a causality loop is formed by the emission and test of S.

```
signal S in
  present S then emit S end
end
```

The computation of signal S is represented by the following formula.

$$S \equiv (\text{START} \wedge S)$$

The statement is semantically incorrect since the status of S cannot be established before the presence test. This translates into the impossibility of proving either S or ¬S assuming START, which corresponds to assuming that the statement is executed. Note that solutions exist in classical logic: both S = true and S = false make the formula true. No solution exists in intuitionistic logic since it would require the proof of S to depend on itself.

Finally, for "present S then emit S else emit S," the equation is

$$S \equiv ((\text{START} \wedge S) \vee (\text{START} \wedge \neg S))$$

If START is true, the equation reduces to $S \equiv S \vee \neg S$, which has no solution in constructive logic, in which the law of excluded middle does not apply.

Ternary circuit simulation

Computing the reaction of an Esterel program can be done by determining the values satisfying the associated constructive logic formula, which in turn can be easily done by representing the formula as a Boolean circuit where the variables become circuit wires and logical operators become gates. Figure 3.2 shows circuits for a pair of intuitionistic formulas. This approach is the basis of the *constructive circuit translation* of Esterel, presented in Chapter 6.

For a circuit, the computation of an instant involves determining the status of every wire based on the status of the inputs. While the status of each input wire is fixed before the circuit is evaluated, all others start off marked with ⊥, which corresponds to when no execution or pruning has been

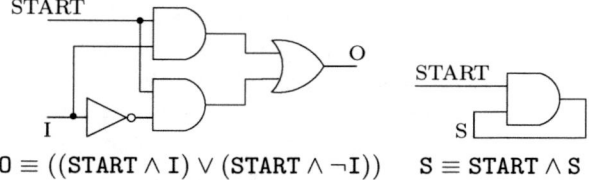

$$O \equiv ((\text{START} \wedge \text{I}) \vee (\text{START} \wedge \neg \text{I})) \qquad S \equiv \text{START} \wedge S$$

Figure 3.2: Boolean circuits for two intuitionistic formulas

x	y	$x \wedge y$	$x \vee y$	$\neg x$
$\bot$	$\bot$	$\bot$	$\bot$	$\bot$
$\bot$	0	0	$\bot$	$\bot$
$\bot$	1	$\bot$	1	$\bot$
0	$\bot$	0	$\bot$	1
0	0	0	0	1
0	1	0	1	1
1	$\bot$	$\bot$	1	0
1	0	0	1	0
1	1	1	1	0

Figure 3.3: Truth table for ternary logic ($\mathcal{B}_\bot$) operators

performed on the Esterel code. The circuit state changes when a gate is given enough input information to compute its output.

For ternary simulation, the computations are all performed in three-valued logic $\mathcal{B}_\bot = \{\bot, 0, 1\}$, with operators $\wedge$, $\vee$, and $\neg$ given in Figure 3.3.

The set $\mathcal{B}_\bot$ is also endowed with a Scott domain structure where $\bot \sqsubseteq 0$ and $\bot \sqsubseteq 1$. Like fact propagation, ternary simulation is monotonic with respect to the order on $\mathcal{B}_\bot$. Once a circuit wire has been set to 0 or 1, i.e., once a new a fact has been determined, the evaluation process cannot change its value. Ternary simulation is a fixpoint computation, which runs until no more wires change their value. The process is finite, and the associated program is correct for the given state and input if every circuit wire has been assigned a final value of 0 or 1.

More information on constructive circuits and ternary simulation will be given in Chapter 6.

Other approaches

The constructive approach is both mathematically well-founded and natural for defining the operational semantics of Esterel. It can be implemented using BDD-based Boolean techniques, using classical dual-rail encoding of ternary logic (see Shiple, Berry, and Touati [65]). However, symbolic causality analysis does not scale to big designs. Also, general constructive causality limits the scope of code analysis and optimization techniques. In particular, widely-accepted optimization techniques in both software and hardware do rely on the law of the excluded middle. Consider the statement

```
present I then emit O else emit O end
```

which might be translated into C as

```
if (I_present)
   O_present = 1;
else
   O_present = 1;
```

At this point, it is tempting to factor out the code fragment "O_present = 1," thus removing the dependency between the test of I and the assignment of O. This is valid for a well-determined input, but not for a not-yet determined local or output signal. In complex cases, this transformation could change the semantics of the statement, i.e., by giving it a behavior instead of causing the program to be rejected. Similar transformations are performed by most digital circuit optimizers.

To scale to big programs and support arbitrary source-code transformations, the simplest way is to require topological acyclicity, which is strictly stronger than constructive causality. This is what is done by production compilers such as Esterel Studio. Similarly, there is little point in generating cyclic circuit since most synthesis systems reject them.

Nevertheless, finding other compromises remains interesting. Several attempts have been made to extend the semantics of Esterel for optimization or proof purposes. Such examples are the Esterel variant Quartz developed by Schneider [62, 61]. Non-constructive transformations are sometimes used in the Saxo-RT Esterel compiler, presented in Section 7.5. We do not present these semantic approaches in this book, but rather explain their implications in the code generation process.

3.2 Flavors of Constructive Semantics

The constructive semantics of Esterel comes in three main flavors that serve different purposes. The *constructive behavioral semantics (CBS)*, defined in Chapter 4, is the simplest and most abstract. It succinctly defines what a program means, and could be used to build interpreters and compilers. However, they would be inefficient. The *constructive operational semantics (COS)*, defined in Chapter 5, is a microstep semantics that analyzes the fine structure of control and signal propagation throughout a reaction and serves as basis for sequential code generation. The *constructive circuit semantics*, defined in Chapter 6, translates an Esterel program into a digital circuit endowed with three-valued logic simulation semantics. The circuit semantics is used to generate actual hardware and to interface to formal verification tools.

While it is generally accepted that the three variants of the constructive semantics are equivalent, the formal equivalence proofs have not been fully completed. It would be very interesting to prove them in formal verification systems such as Coq [15] or HOL [36], used by Schneider [60, 59] to prove the correctness of its Quartz variant of Esterel.

Constructive Behavioral Semantics

The main differences among the three semantics come in how they encode control flow and the propagation of facts. Control flow can be viewed as a form of fact propagation.

The constructive behavioral semantics, due to Berry [7], takes this approach and computes the reaction of a statement or program by mainly using recursive fact-propagation routines called *potential functions*.

- The *Must* function encodes control flow propagation. It recursively determines which statements are executed and which signals are emitted by a statement given the status of each input signal and the current knowledge about control and signal facts.

- The *Can* function performs code pruning. Given a statement and a current facts about signals, it prunes inaccessible statements and reports which signals can be declared absent.

- The Can^+ function combines the control flow and code pruning into a single potential function. Given a statement and a signal context, it reports all the signals that must be emitted or can be emitted knowing that the statement will be executed.

In Berry [7], *Can* and Can^+ were grouped in a single function Can^m to factor out rules. Here, we split the definitions so we can reuse and extend *Can* in the COS.

The final output behavior of the program is determined using recursive calls to *Must*, *Can*, and Can^+. The correctness of the program is also determined in the process: a program is rejected as incorrect for the given state and signal context if the calls leave the status of any signal unknown.

Once the input/output behavior of the program has been determined, a second step computes the resulting program state. This part of the semantics is given in the inference-rule style of Plotkin's [54] structural operational semantics.

Relying on potential functions for the representation of control flow leads to relatively few inference rules. The main inconvenience is that the macro-step approach is unable to discern intra-instant causality and concurrency, so that dealing with data is problematic and hence using the CBS as a basis for code generation does not appear practical.

Constructive Operational Semantics

To generate fast sequential C code from Esterel programs, we wish to execute only active statements in each reaction. To this end, the constructive operational semantics represents statement execution using classical control flow notation instead of a *Must* potential function. Transition rules represent the elementary execution and synchronization steps that form the computation of an instant. For determining signal absence through code pruning, it still relies on the *Can* potential function, but control flow and code pruning can be separated, so that the combined Can^+ is no longer necessary.

The V3 semantics introduced by Berry and Gonthier [10] was the first operational semantics for Esterel. That semantics is a subset of the current constructive semantics, based on more restrictive and more complex potential analysis techniques. Its determinism was formally proven by Gonthier. However, once the current simpler and more scalable constructive approach was adopted as the basis for Esterel, a new operational semantics was needed. The constructive operational semantics, due to Potop-Butucaru [56], is fully compatible with the constructive behavioral semantics. It has the fine grain needed to represent the causality and concurrency of data computations and therefore allows the definition of the semantics of data-handling Esterel programs.

The COS can be considered as the link between the CBS and code generation. On one hand, it is a refinement of the CBS based on a more direct presentation of *Must* by control-flow propagation using microsteps. To determine the reaction of an instant or the non-constructiveness of a program, microsteps are chained into executions. On the other hand, we shall see in Chapter 8 how the control structures of the COS directly correspond to the different elements of the GRC intermediate format, upon which efficient Esterel compilers are based.

To facilitate understanding and to introduce concepts and structures in a gradual way, we present the constructive behavioral semantics in Chapter 4 and the COS in Chapter 5. The CBS first defines computation cycles, reactivity and determinism, which form the minimal requirements any other semantics must obey. Then, the COS refines the notation and the transition rules to give an operational way of computing them.

Constructive Circuit Semantics

In Section 3.1.1, we explained that the semantics of Esterel can be defined by representing programs with constructive logic formulas. Ternary simulation of the associated circuit representation can be used to determine the correctness and the reaction of a program for given state and input. This approach forms the basis of the *constructive circuit semantics* of Esterel, where Esterel programs generate Boolean circuits. At the circuit level, semantics and correctness are defined by constructive logic rules or equivalently by ternary simulation. Shiple, Berry, and Touati [65] proved the equivalence between the three-valued simulation semantics and the up-bounded inertial delay model due to Brzozowski and Seger [18], a realistic, but very conservative model of digital circuits.

In practice, the circuit semantics is used by the Esterel Studio compiler to generate efficient circuits from Esterel V7 programs. Although the language has richer data-manipulation constructs, compiling control structures is like that in V5. For hardware applications, generated circuits are usually required to be acyclic. They can be fed into hardware synthesis systems and model checkers for formal verification. For acyclic circuits, binary simulation suffices. The circuit semantics are also used by the INRIA and Esterel Studio compilers to generate C. The C code is essentially a C representation of the circuit gates, topologically sorted to guarantee that dependencies are respected during the evaluation. Ternary circuit simulation has also been used to compile Esterel programs to C. The INRIA compiler includes a ternary simulator to evaluate the reaction of a constructive cyclic circuit to an input by fact propagation. For such C code generation from either cyclic or acyclic circuits generated from Esterel, there is an intrinsic inefficiency: all gates must be evaluated, including the ones that do not affect outputs. Evaluation time is linear in the size of the program, mostly irrespective of behavior.

The INRIA Esterel compiler also includes a BDD-based ternary model checker that tests for constructiveness and replaces the cyclic circuit by a behaviorally equivalent acyclic one. It can be used for hardware or software generation, but it does not scale well to very large programs.

The circuit semantics presented here, due to Potop-Butucaru, is slightly different from the original circuit semantics of Berry [7]. Potop-Butucaru's thesis [56] gives a detailed comparison and proves their equivalence.

3.3 Conventions and Preliminary Definitions

3.3.1 Global Correctness of an Esterel Program

The various formal semantics of Esterel, including the ones presented in this book, concentrate on defining the behavior and correctness (constructivity) of a program *for a given state and status of input signals.*

We say that a program is *globally constructive* if the program is correct for every reachable state and for any input signal combination compatible with the signal relations specified in the program interface.

In the third part of the book, we shall show that checking the global correctness of a program is a complex issue unless dependencies are acyclic.

3.3.2 Restriction to Kernel Esterel

In defining the semantics of Esterel with data, we restrict ourselves to the Kernel Esterel language, defined in Section 2.2. The derived statements get their semantics as macro definitions over the kernel language. The same approach is used in the circuit translation definition, where direct translation rules are only given for the kernel language primitives. However, actual compilers use more efficient rules for certain derived statements.

3.3.3 Signal Events

To represent the status of each signal visible to a statement, we rely on *signal events*. Given a set Σ of signals, also called a *sort*, an event E over Σ associates a status $E(S) \in \mathcal{B}_\perp = \{\perp, 0, 1\}$ with every signal $S \in \Sigma$. The sort of E is denoted Sort(E). The valuation $E(S) = 1$ corresponds to S being present, while $E(S) = 0$ corresponds to S being absent. The valuation $E(S) = \perp$ corresponds to the case where the status of S has not yet been determined by the constructive semantics. In this case, we say that the status of S is *unknown*.

To improve the readability of the semantic rules, we use a set-like notation for events. The notation ignores the sort and the unknown signal statuses. Thus, $\{S^1, P^1, Q^0\}$ represents the event where S and P are present, Q is absent, and all the other signals have $\perp$ status. Of course, $\{S^1, S^0\}$ is not an acceptable signal event, since a signal cannot have status 0 and 1 at the same time. We write $S^m \in E$ if $E(S) = m$ and we denote with $\emptyset$ the event where all signals have unknown status. The set-like notation also supports the intuition that events of support Σ are also events of support $\Sigma' \supseteq \Sigma$, the extra signals having status unknown in the larger support. The set of events of a given sort is partially ordered by $\leq$ in a pointwise way on signals: $E \leq E'$ means $E(S) \leq E'(S)$ for all S. Note that $\leq$ is the partial order induced by the partial order on the $\mathcal{B}_\perp$ domain.

Given a signal set Σ, a signal $S \in \Sigma$, and E an event of sort Σ, we write $E \setminus S$ for the event of sort $\Sigma \setminus \{S\}$ that coincides with E on its sort. Given an event E of sort Σ, a signal S, and $m \in \mathcal{B}_\perp$, we denote by $E * S^m$ the event of sort $\Sigma \cup \{S\}$ with $E * S^m(S) = m$ and $E * S^m(Q) = E(Q)$ for all $Q \neq S$. Notice that the status of S in E is lost in $E * S^m$ if $S \in \text{Sort}(E)$. This means that declarations of identically-named signals can be safely nested without a stack, e.g., "**signal** S **in signal** S **in** p **end end**."

3.3.4 Trap Handling and Completion Codes

First introduced by Gonthier [35] as part of the Esterel V3 semantics, the integer encoding of statement completion status is key to understanding the semantics and generating efficient code. The idea is to assign a distinct integer to each way a statement may terminate to deal with the interaction between traps and parallel synchronization.

An Esterel statement that is started or resumed during an execution instant completes its execution instant by doing exactly one of the following:

- *terminate* its execution normally and pass on the control to the next statement in sequence;

- *pause*, that is retain the control flow in one or more **pause** sub-statements (so that it can be resumed in a subsequent instant), without exiting any trap; or

- *exit* a trap. In case several traps are exited simultaneously, only the highest-level one is reported to the enclosing statement.

The completion codes of statements running in parallel need to be composed; the integer codes have been chosen to make this a simple "max" operation.

- 0 is the code for normal termination;

- 1 is the code for pause; and

- $k+2$ is the code associated with trap T, if k other **trap** declarations have to be traversed before reaching the declaration of T.

Consider the statement in Figure 3.4 in the instant where it starts. The completion code of "**nothing**" is 0, as the statement instantly terminates; "**pause**" completes with code 1, as it retains control; and "**exit T**" produces code 2, as the declaration of T is the first enclosing trap declaration. The two "**exit U**" statements produce different completion codes. The first is separated from the declaration of U by the declaration of T. Therefore, it exits with code 3. The second instance of "**exit U**" exits with code 2.

The parallel statement takes the maximum of the completion code of all its branches, which by design is exactly all the synchronization it is required to do. Consider a parallel statement $p \,||\, q$, and assume that p returns k and q returns l. There are three possibilities:

- the parallel terminates if and when both branches have terminated, since $\max(k,l) = 0$ is equivalent to $k = l = 0$;

- the parallel pauses if one branch pauses and the other one does not exit a trap, since $\max(k,l) = 1$ implies $k = 1$ and $l \leq 1$ or vice-versa; or

- the parallel exits a trap if one branch does. If both branches exit traps, the parallel only propagates the outermost one. This follows from the encoding of exit statements that starts with code 2 and increases by 1 for each enclosing trap.

In our example, the four-branch parallel completes with code 3 in the first instant.

```
trap U in                % code 0
    trap T in            % code 2
        nothing          % code 0
        ||               % code 3
        pause            % code 1
        ||
        exit T           % code 2
        ||
        exit U           % code 3
    end
    ||                   % code 2
    exit U               % code 2
end
;                        % code 1
pause                    % code 1
```

Figure 3.4: Completion codes in the start instant of a simple statement

When the body of a trap declaration terminates or pauses (completion codes 0 or 1), the trap declaration itself does the same. The trap declaration statement is also the handler of the trap it defines. It terminates when it receives the completion code 2 from its body, which corresponds to the body exiting the declared trap. In this case, the trap declaration preempts its body and terminates. The body producing a completion code $k \geq 3$ corresponds to a higher-level trap being exited. To report the exit to the environment, the trap declaration completes with code $k - 1$. The decrement corresponds here to the traversal of one trap declaration (which is consistent with the definition of the completion codes).

In our example, the declaration of "trap T" completes with code 2, as its body completes with code 3. Then the innermost parallel statement completes with code 2 and the trap is handled by the declaration of U, whose body is preempted. Therefore, that the first "pause" statement does not retain control between instants. Control is passed to the next instruction in sequence and it pauses in the second "pause" statement. Consequently the top-level sequence statement pauses (code 1).

If the statement is resumed in the following execution instant, the second "pause" statement terminates (code 0), and the entire sequence terminates. The other statements produce no codes since they do not execute.

Program completion

As explained in Section 2.5, the body statement of a module can have no free trap. Therefore, a program can complete only with code 0, representing program termination, or 1, which tells us that the program paused (i.e., that it did not terminate and will be resumed in the future). The reaction function produced by Esterel compilers returns this code to the environment.

Notation

Certain choices of notation prove useful in defining the semantics of Esterel. We shall annote the statement "exit T" with a completion code, e.g., "exit T(k)", where $k \notin \{0, 1\}$ is the completion code associated with T at the given place in the Esterel program.

Parallel synchronization is represented by the max operator; we need to extend it to sets of completion codes.

$$\max(K, L) = \begin{cases} \emptyset & \text{if } K = \emptyset \text{ or } L = \emptyset \\ \{\max(k, l) : k \in K, l \in L\} & \text{if } K, L \neq \emptyset \end{cases}$$

The downshift operator ($\downarrow$) decrements trap codes when they traverse a trap declaration. Again, we define the operator on single values

$$\downarrow k = \begin{cases} 0 & \text{if } k \in \{0, 2\} \\ 1 & \text{if } k = 1 \\ k - 1 & \text{otherwise} \end{cases}$$

and sets

$$\downarrow K = \{\downarrow k \mid k \in K\}.$$

4

Constructive Behavioral Semantics

In this chapter, we define the semantic terms representing the state of an Esterel statement; analyze potentials, including the full set of rules for computing the three potential functions; and give the structural induction rules used to compute the behavioral transitions. The following sections explain how program transitions are derived from those of its body, and investigate some correctness issues.

4.1 Behavioral Transitions

4.1.1 Transition syntax

The CBS formalize a reaction of a program P as a macro-step *behavioral transition* of the form
$$P' \xrightarrow[I]{O} P''.$$
Here, P' and P'' are decorated program texts that represent possibly identical successive execution states of P. Representing program states by program text is standard for rules in structural operational semantics rules [54, 52].

The *input event* I gives the present/absent status of each input signal (provided by the environment). The *output event* O gives the status of each output signal as computed by the program.

The actual computation of the reaction is done using an auxiliary *statement transition* relation, which has the form
$$\overline{p} \xrightarrow[E]{E', k} \overline{p}',$$
where p is a statement, $\overline{p}$ and $\overline{p}'$ are terms defining the *state* of p before and after the transition, E is the input signal event of the transition, E' is the

output event of the transition, and k is the integer *completion code* returned by p following the definition in Chapter 3.

The two transition systems are related as follows: p and p' in the second relation are states of the body of P, with the interface stripped away. To handle output signals of P given I, the input environment E for the body transition must be computed with care because P's outputs are also fed back to the body. This will be explained in Section 4.7.

As we explained in the last chapter, the computation of the statement transition relation for a given statement, state, and signal event is performed in two steps. First, the *analysis of potentials* uses the functions *Must*, *Can*, and Can^+ to determine the correctness of the statement, its output event, and its completion code. In the second step, the full statement transition relation, which also defines the resulting program state, is computed using structural induction rules.

4.1.2 States as Decorated Terms

Consider the term $p = $ pause ; pause ; emit S. In the CBS, the idea is to keep the shape of the statement constant and to decorate it to indicate where control pauses between reactions. Only occurrences of pause are decorated:* we write $\widehat{\text{pause}}$ instead of pause to indicate that execution has paused and will resume from there in the following instant. The other statements do not need to be decorated since they are instantaneous. Using these decorations, the execution—the sequence of reactions—of p is

$$\text{pause ; pause ; emit S} \xrightarrow[\emptyset]{\emptyset, 1} \widehat{\text{pause}} \text{ ; pause ; emit S}$$

$$\xrightarrow[\emptyset]{\emptyset, 1} \text{pause ; } \widehat{\text{pause}} \text{ ; emit S}$$

$$\xrightarrow[\emptyset]{\{S\}, 0} \text{pause ; pause ; emit S}$$

The boot statement

There is a slight problem if we assume the previous statement is the entire body of a program. Indeed, in the above example, the fourth term is $p_3 = $ pause ; pause ; emit S, which is the same as the initial term p_0. However, the program should be considered terminated in p_3 but not in p_0. Using such a rewriting would assume the environment in which the program runs differentiates between the "terminated" state and "not yet started."

All current approaches to compiling Esterel instead place the burden on the program itself. This done by replacing the initial statement p_0 with the decorated term "$\widehat{\text{pause}}$; p_0." Then, control initially resumes from the first

---

*In the full language, other statements can be decorated, for example await, every, and loop-each. The expansion of each of these user-level statements involves exactly one pause statement, which makes decoration unambiguous.

$\widehat{\texttt{pause}}$ statement, called the *boot* statement. An undecorated program such as p_3 is considered terminated. In this approach, the extended program term is never started, but only resumed. With the boot $\widehat{\texttt{pause}}$ statement added, the execution sequence becomes

$$\widehat{\texttt{pause}} \;;\; \texttt{pause} \;;\; \texttt{pause} \;;\; \texttt{emit S} \quad \xrightarrow[\emptyset]{\emptyset,\,1} \quad \texttt{pause} \;;\; \widehat{\texttt{pause}} \;;\; \texttt{pause} \;;\; \texttt{emit S}$$

$$\xrightarrow[\emptyset]{\emptyset,\,1} \quad \texttt{pause} \;;\; \texttt{pause} \;;\; \widehat{\texttt{pause}} \;;\; \texttt{emit S}$$

$$\xrightarrow[\emptyset]{\{S\},\,0} \quad \texttt{pause} \;;\; \texttt{pause} \;;\; \texttt{pause} \;;\; \texttt{emit S}$$

We could have used a termination marker instead of a boot statement, but the boot statement is more natural in the circuit translation and avoids the need for a new symbol.

4.1.3 State Syntax

In the definition of the semantics, the letters p and q represent Kernel Esterel statements. To simplify the definition of the semantic rules, we generalize the decoration of **pause** statements to all kernel statements. I.e., $\widehat{p}$ is a state of p where one or more of its **pause** statements are decorated. Here, we also say that the statement p is *selected for execution in the next reaction*, or simply *selected*. We write $\overline{p}$ to indicate a term over p that may or may not be selected,. In other words, $\overline{p}$ represents either a selected term $\widehat{p}$, or the unselected term p. The grammar of states and terms is

$$\begin{aligned}
\widehat{p} \;::=\;& \widehat{\texttt{pause}} \\
\mid\;& \texttt{present } S \texttt{ then } \widehat{p} \texttt{ else } p \texttt{ end} \\
\mid\;& \texttt{present } S \texttt{ then } p \texttt{ else } \widehat{p} \texttt{ end} \\
\mid\;& \texttt{suspend } S \texttt{ when } \widehat{p} \\
\mid\;& \widehat{p}\,;\,p \\
\mid\;& p\,;\,\widehat{p} \\
\mid\;& \texttt{loop } \widehat{p} \texttt{ end} \\
\mid\;& \widehat{p} \parallel \overline{p} \\
\mid\;& \overline{p} \parallel \widehat{p} \\
\mid\;& \texttt{trap } T \texttt{ in } \widehat{p} \texttt{ end} \\
\mid\;& \texttt{signal } S \texttt{ in } \widehat{p} \texttt{ end} \\
\overline{p} \;::=\;& p \\
\mid\;& \widehat{p}
\end{aligned}$$

Notice that a subterm is selected if and only if it contains a selected **pause** statement $\widehat{\texttt{pause}}$. In a selected term such as **present** S **then** $\widehat{p}$ **else** q **end**, the only selected subterm is $\widehat{p}$. Such a state is reached when the test has taken its left branch in some previous instant and execution of this branch has not yet terminated (and has not been preempted). There is no term of the form **present** S **then** $\widehat{p}$ **else** $\widehat{q}$ **end** because one cannot be in the **then** and `else` branches at the same time. Similarly, there is no term of the form $\widehat{p}\,;\,\widehat{q}$ since it is not possible to pause in both components of a sequence at the same time.

A branch of a selected parallel statement may or may not be selected. In $\widehat{p} \parallel \widehat{q}$, both $\widehat{p}$ and $\widehat{q}$ are selected: the parallel has been started in some past instant and both branches are still active. A term of the form $\widehat{p} \parallel q$ represents the case where the second branch of the parallel has terminated while the first one is still active. For example, the term $(\widehat{\text{pause}}\,;\,\text{pause}) \parallel \widehat{\text{pause}}$ becomes $(\text{pause}\,;\,\widehat{\text{pause}}) \parallel \text{pause}$ in the following instant.

4.2 Analysis of Potentials

Consider a program P with body p in state $\widehat{p}$, and let I be an input event that associates a present/absent/undefined status to every input signal. As explained in Chapter 3, the constructive behavioral semantics determines the status of the output signals and the completion code of the program through calls to the potential functions *Must*, *Can*, and Can^+. The computation of the functions is performed through recursive calls that follow the control flow structure of the program.

Execution—The Must Function

The *Must* function determines which signals will be emitted when executing a statement in a given signal context. If the execution of the program does not block for the given input event and program state, the *Must* function also determines the completion code of p. The *Must* function has the form

$$Must(\overline{p}, E) = \langle\, F\,,\, K\,\rangle.$$

Here, E is the signal context—a signal event that associates a status (1, 0, or $\bot$) with every input or output signal of p. The signal event F gives the set of signals that p must emit. It only assigns statuses 1 and $\bot$. The set K is empty if the execution of $\widehat{p}$ blocks for the input event E. If execution completes with code k, then $K = \{k\}$. To independently manipulate the components of the result pair, we identify them using subscripts:

$$Must(\overline{p}, E) = \langle\, Must_S(\overline{p}, E)\,,\, Must_K(\overline{p}, E)\,\rangle.$$

Code Pruning—The Can Function

The function $Can(\overline{p}, E)$ is used to prune impossible paths and determine which signals and completion codes could possibly be produced after pruning. Its form is similar to that of *Must*:

$$Can(\overline{p}, E) = \langle\, Can_S(\overline{p}, E)\,,\, Can_K(\overline{p}, E)\,\rangle.$$

Here, $Can_S(\overline{p}, E)$ is the set of emit statements that have not been invalidated by the given signal context E^*. Similarly, $Can_K(\overline{p}, E)$ computes the set of not-invalidated completion codes.

Combined Potential—The Can^+ Function

The *Must* function only represents execution and the *Can* function only represents code pruning over not-yet-started statements. However, it is often the case that the execution blocks inside a statement and we still need to determine what the not-yet-executed code can do. To reason about such cases, we use the Can^+ function, which combines execution and code pruning. The form of Can^+ is similar to that of the other potential functions:

$$Can^+(\overline{p}, E) = \langle\, Can_S^+(\overline{p}, E)\,,\, Can_K^+(\overline{p}, E)\,\rangle.$$

Here, $Can_S^+(\overline{p}, E)$ includes all the signals that must be emitted by p and all the signals that can be emitted after the execution is blocked if it does.

Notation

We extend the inclusion predicate $\subseteq$ and the union operator $\cup$ componentwise on pairs $\langle\, F,\, K\,\rangle$. We also extend the signal restriction operator to pairs:

$$\langle\, F,\, K\,\rangle \setminus S =_{def} \langle\, F \setminus S,\, K\,\rangle.$$

For presentation reasons, we shall use the vertical and horizontal pair notation interchangeably when representing potential functions:

$$\langle\, F,\, K\,\rangle = \left\langle\, \begin{matrix} F \\ K \end{matrix} \,\right\rangle.$$

Example

Consider the statement p from Figure 3.1:

```
signal R in
  emit S ;
  present R then emit O1 else emit O2 end ;
  present S else emit R end
end
```

Assume the execution of p starts in a context where the signal S is already present (i.e., emitted by the environment). The first step is to prune unreachable code. Here, because the **else** branch of "**present S**" cannot be taken,

---

*Section 4.6 shows how upper bounds for $Can_K(p, E)$ can be defined for use in the current code generation schemes.

signal R cannot be emitted and hence cannot be present. This implies that O1 cannot be emitted. Therefore

$$Can(p, E) = \langle\, \{S^1, O2^1\}\,,\, \{0\}\,\rangle.$$

Since R cannot be emitted ($R^1 \notin Can_S(p, E)$), we can set it absent (0). Execution proceeds by emitting S, and performing the two signal tests, which results in O2 being emitted. Therefore,

$$Must(p, E) = \langle\, \{S^1, O2^1\}\,,\, \{0\}\,\rangle.$$

Since the execution does not block, we also have $Can^+(p, E) = Must(p, E)$.

Now assume the execution of p starts in a context F where S has not been emitted by the environment. Since the status of S is still unknown ($\bot$), no code can be pruned and we have

$$Can(p, F) = \langle\, \{S^1, T^1, O1^1, O2^1\}\,,\, \{0\}\,\rangle.$$

The control flow simulation, i.e., the *Must* computation, starts by setting the status of R to unknown ($\bot$). Then, "emit S" is executed, S is emitted and control flow blocks on the first signal test. We have

$$Must(p, F) = \langle\, \{S^1\}\,,\, \emptyset\,\rangle.$$

The Can^+ potential, which combines execution and code pruning, is

$$Can^+(p, F) = \langle\, \{S^1, T^1, O1^1, O2^1\}\,,\, \{0\}\,\rangle.$$

The first computation of the potential functions is completed, but not the potential computation step. Indeed, we determined new information: the fact that S is emitted in the context F. We need to feed this information back by adding it to the input environment and recomputing the potential functions. Notice that the new environment $E = F \cup \{S^1\}$ corresponds to the first case we considered.

It is important to note how the potential computation advances by successively enriching the environment of each statement. In our example, in the second case the initial call with environment F is followed by a call with environment $F \cup \{S^1\}$. Potential computation stops when no new information can be determined through calls to the potential functions. The monotonicity property is essential for ensuring the convergence of the computation.

4.2.1 The Definition of Must, Can, and Can$^+$

This section defines the three potential functions for the kernel language statements. We first define the potential functions for non-decorated terms, used when a statement is started. Then, we reduce the potential computation over decorated terms (corresponding to statement resumption) to the non-decorated case.

Non-decorated terms

The definitions of the three functions coincide for signal emissions and for basic completion code generators.

$$
\begin{aligned}
Must(\text{nothing}, E) &= Can(\text{nothing}, E) = Can^+(\text{nothing}, E) = \langle \emptyset, \{0\} \rangle \\
Must(\text{pause}, E) &= Can(\text{pause}, E) = Can^+(\text{pause}, E) = \langle \emptyset, \{1\} \rangle \\
Must(\text{exit T}(k), E) &= Can(\text{exit T}(k), E) = Can^+(\text{exit T}(k), E) = \langle \emptyset, \{k\} \rangle \\
Must(\text{emit S}, E) &= Can(\text{emit S}, E) = Can^+(\text{emit S}, E) = \langle \{S^1\}, \{0\} \rangle
\end{aligned}
$$

Now, consider a signal test "**present** S **then** p **else** q **end**" and an event E. First, for both *Must* and *Can*, the definitions are easy if the status of S in E is either 1 or 0: we recursively analyze the first branch if the status is 1 and the second branch if the status is 0. If the status of S in E is unknown, *Must* and *Can* behave very differently. For *Must*, we return the empty signal set and the empty completion code set since none of the branches must be taken. For *Can*, we return the union of the sets of signals that the branches can emit and the union of the sets of completion codes the branches can return. Since nothing is executed, Can^+ behaves like Can:

$$
Must(\textbf{present } S \textbf{ then } p \textbf{ else } q \textbf{ end}, E) = \begin{cases} Must(p, E) & \text{if } S^1 \in E \\ Must(q, E) & \text{if } S^0 \in E \\ \langle \emptyset, \emptyset \rangle & \text{if } S^\perp \in E \end{cases}
$$

$$
Can(\textbf{present } S \textbf{ then } p \textbf{ else } q \textbf{ end}, E) = \begin{cases} Can(p, E) & \text{if } S^1 \in E \\ Can(q, E) & \text{if } S^0 \in E \\ Can(p, E) \cup Can(q, E) & \text{if } S^\perp \in E \end{cases}
$$

$$
Can^+(\textbf{present } S \textbf{ then } p \textbf{ else } q \textbf{ end}, E) = \begin{cases} Can^+(p, E) & \text{if } S^1 \in E \\ Can^+(q, E) & \text{if } S^0 \in E \\ Can(p, E) \cup Can(q, E) & \text{if } S^\perp \in E \end{cases}
$$

When started, a suspension statement acts like its body:

$$
\begin{aligned}
Must(\textbf{suspend } S \textbf{ when } p, E) &= Must(p, E) \\
Can(\textbf{suspend } S \textbf{ when } p, E) &= Can(p, E) \\
Can^+(\textbf{suspend } S \textbf{ when } p, E) &= Can^+(p, E).
\end{aligned}
$$

For a sequence $p\,;\,q$, we analyze q only if p must (resp. can) terminate, in which case the completion code 0 of p is discarded. Note the complexity of the Can^+ definition, whose cases correspond to cases of both *Must* and *Can*:

$Must(p\ ;\ q, E) =$
$$\begin{cases} Must(p, E) & \text{if } 0 \notin Must_K(p, E), \\ \left\langle \begin{array}{l} Must_S(p, E) \cup Must_S(q, E) \\ Must_K(q, E) \end{array} \right\rangle & \text{otherwise.} \end{cases}$$

$Can(p\ ;\ q, E) =$
$$\begin{cases} Can(p, E) & \text{if } 0 \notin Can_K(p, E), \\ \left\langle \begin{array}{l} Can_S(p, E) \cup Can_S(q, E) \\ Can_K(p, E) \setminus \{0\} \cup Can_K(q, E) \end{array} \right\rangle & \text{otherwise.} \end{cases}$$

$Can^+(p\ ;\ q, E) =$
$$\begin{cases} Can^+(p, E) & \text{if } 0 \notin Can_K^+(p, E), \\ \left\langle \begin{array}{l} Can_S^+(p, E) \cup Can_S^+(q, E) \\ Can_K^+(p, E) \setminus \{0\} \cup Can_K^+(q, E) \end{array} \right\rangle & \begin{array}{l}\text{if } 0 \in Can_K^+(p, E) \\ \text{and } 0 \in Must_K(p, E),\end{array} \\ \left\langle \begin{array}{l} Can_S^+(p, E) \cup Can_S(q, E) \\ Can_K^+(p, E) \setminus \{0\} \cup Can_K(q, E) \end{array} \right\rangle & \text{otherwise.} \end{cases}$$

The result of $Must_K(p, E)$ is either empty or a singleton, meaning execution is deterministic. Note that in the definition of $Must(p\ ;\ q, E)$, we could also write the predicates using set comparisons, i.e., as $Must_K(p, E) \neq \{0\}$ and $Must_K(p, E) = \{0\}$.

Note that in the semantics of $p\ ;\ q$, the signals emitted by p are not fed into q at the level of the sequence potential functions. This could be done, but we prefer obtaining the same semantics while maintaining a strict separation of concerns. In our framework, signal feedback is only performed at the level of signal declaration statements. The global feedback effect is obtained by performing successive calls of the potential functions, with increasing environments. See the handling of signal declaration below.

A loop is an infinite sequence of occurrences of the loop body. However, the body cannot be instantaneous, so that only its first instance must be traversed by the potential functions. The fact that the body cannot terminate will be ensured by the inference rules over transitions.

$$\begin{aligned} Must(\texttt{loop}\ p\ \texttt{end}, E) &= Must(p, E) \\ Can(\texttt{loop}\ p\ \texttt{end}, E) &= Can(p, E) \\ Can^+(\texttt{loop}\ p\ \texttt{end}, E) &= Can^+(p, E) \end{aligned}$$

The signal potential of a parallel is the union of the signal potentials of the branches. The completion code potential is obtained by combining the potentials of the branches using the max operator, as explained in Section 3.3.4:

$$Must(p \parallel q, E) = \left\langle \begin{array}{c} Must_S(p, E) \cup Must_S(q, E) \\ \max(Must_K(p, E), Must_K(q, E)) \end{array} \right\rangle$$

$$Can(p \parallel q, E) = \left\langle \begin{array}{c} Can_S(p, E) \cup Can_S(q, E) \\ \max(Can_K(p, E), Can_K(q, E)) \end{array} \right\rangle$$

$$Can^+(p \parallel q, E) = \left\langle \begin{array}{c} Can_S^+(p, E) \cup Can_S^+(q, E) \\ \max(Can_K^+(p, E), Can_K^+(q, E)) \end{array} \right\rangle$$

Note that the completion codes of both branches are required to compute the completion code of the parallel. In other terms, $Must_K(p \parallel q, E)$ is non-empty, and therefore a singleton set, if and only if $Must_K(p, E)$ and $Must_K(q, E)$ are non-empty.

Using the max set operation in the definition of control propagation ensures that a parallel cannot terminate if one of its branches cannot. This property would be lost if max were replaced by a simple union, and this would lead to abnormally rejecting constructively correct programs such as

```
[
    present S then emit O end
||
    pause
] ; emit S
```

where the parallel statement pauses in the first instant because of its second **pause** branch, breaking the potential cycle on S.

For traps, we apply the appropriate operators, defined in Section 3.3.4, to the completion codes returned by the body:

$$Must(\textbf{trap } T \textbf{ in } p \textbf{ end}, E) = \langle \, Must_S(p, E) \, , \, \downarrow Must_K(p, E) \, \rangle$$
$$Can(\textbf{trap } T \textbf{ in } p \textbf{ end}, E) = \langle \, Can_S(p, E) \, , \, \downarrow Can_K(p, E) \, \rangle$$
$$Can^+(\textbf{trap } T \textbf{ in } p \textbf{ end}, E) = \langle \, Can_S^+(p, E) \, , \, \downarrow Can_K^+(p, E) \, \rangle.$$

Signal rules for non-decorated terms

The rules for the local signal declaration operator "**signal** S **in** p **end**" are very different for *Must* and *Can*. For *Must*, because of the way the recursion works, the rule is used only if we already know that "**signal** S **in** p **end**" must be executed. Since we have no information yet about the status of S, we first set this status to $\bot$ and we compute what we must and cannot do. If we find that p must emit S, we take this fact for granted and re-analyze p with status 1 for S. If we find that p cannot emit S, we know that S must be absent and we re-analyze p with status 0 for S. Otherwise, we cannot progress. In each case, we enforce the scoping rule by removing the status of the local S from the emitted signal set:

$$Must(\texttt{signal } S \texttt{ in } p \texttt{ end}, E) = \begin{cases} Must(p, E * S^1) \setminus S & \text{if } S^1 \in Must_S(p, E * S^\perp), \\ Must(p, E * S^0) \setminus S & \text{if } S^1 \notin Can_S^+(p, E * S^\perp), \\ Must(p, E * S^\perp) \setminus S & \text{otherwise}. \end{cases}$$

For Can, we first analyze the body p with the status of S set to $\perp$. If the signal cannot be emitted, we can consider the signal absent and we re-analyze the statement with the status of S set to 0. If the signal cannot be considered absent, we simply return the result of the first analysis:

$$Can(\texttt{signal } S \texttt{ in } p \texttt{ end}, E) = \begin{cases} Can(p, E * S^0) \setminus S & \text{if } S^1 \notin Can_S(p, E * S^\perp), \\ Can(p, E * S^\perp) \setminus S & \text{otherwise} \end{cases}$$

The combined potential function Can^+ combines the analysis of $Must$ and Can:

$$Can^+(\texttt{signal } S \texttt{ in } p \texttt{ end}, E) = \begin{cases} Can^+(p, E * S^1) \setminus S & \text{if } S^1 \in Must_S(p, E * S^\perp) \\ Can^+(p, E * S^0) \setminus S & \text{if } S^1 \notin Can_S^+(p, E * S^\perp) \\ Can^+(p, E * S^\perp) \setminus S & \text{otherwise} \end{cases}$$

The last rules make the difference between Can and Can^+ explicit. The latter is allowed to recursively call $Must$ to represent execution, whereas Can is just a pruning function that can only call itself.

Extension to state terms

When applying $Must$ and Can to decorated state terms, i.e., to statements that held control from the previous instant, the computation must be driven towards the active sub-statements. The rules effectively isolate the non-decorated terms representing the transition function of the current reaction.

A selected **pause** statement behaves like **nothing**. Regardless of the environment, the statement terminates and no signal is produced:

$$Must(\widehat{\texttt{pause}}, E) = Can(\widehat{\texttt{pause}}, E) = Can^+(\widehat{\texttt{pause}}, E) = \langle \emptyset, \{0\} \rangle.$$

When a selected **present** statement is resumed, the test signal is ignored and the statement behaves like its selected branch. Therefore, the potentials of the **present** statement are the potentials of the selected sub-statement:

$$\begin{aligned} Must(\texttt{present } S \texttt{ then } \widehat{p} \texttt{ else } q \texttt{ end}, E) &= Must(\widehat{p}, E) \\ Can(\texttt{present } S \texttt{ then } \widehat{p} \texttt{ else } q \texttt{ end}, E) &= Can(\widehat{p}, E) \\ Can^+(\texttt{present } S \texttt{ then } \widehat{p} \texttt{ else } q \texttt{ end}, E) &= Can^+(\widehat{p}, E) \end{aligned}$$

$$Must(\text{present } S \text{ then } p \text{ else } \widehat{q} \text{ end}, E) = Must(\widehat{q}, E)$$
$$Can(\text{present } S \text{ then } p \text{ else } \widehat{q} \text{ end}, E) = Can(\widehat{q}, E)$$
$$Can^+(\text{present } S \text{ then } p \text{ else } \widehat{q} \text{ end}, E) = Can^+(\widehat{q}, E).$$

If the test signal is present in the environment ($S^1 \in E$), a selected **suspend** statement freezes the state of its body and pauses (completion code 1). No signal is emitted. In this case, all three potential functions return the pair $\langle \emptyset, \{1\} \rangle$. When the test signal is absent ($S^0 \in E$), the **suspend** statement behaves like its body. When the status of S is not yet known ($S^\perp \in E$), no execution takes place, so the $Must$ function returns the empty pair. However, the body still can be executed or preempted. Therefore, Can and Can^+ return the union of the potentials of the first two cases:

$$Must(\text{suspend } S \text{ when } \widehat{p}, E) = \begin{cases} \langle \emptyset, \{1\} \rangle & \text{if } S^1 \in E \\ Must(\widehat{p}, E) & \text{if } S^0 \in E \\ \langle \emptyset, \emptyset \rangle & \text{if } S^\perp \in E \end{cases}$$

$$Can(\text{suspend } S \text{ when } \widehat{p}, E) = \begin{cases} \langle \emptyset, \{1\} \rangle & \text{if } S^1 \in E \\ Can(\widehat{p}, E) & \text{if } S^0 \in E \\ \langle \emptyset, \{1\} \rangle \cup Can(\widehat{p}, E) & \text{if } S^\perp \in E \end{cases}$$

$$Can^+(\text{suspend } S \text{ when } \widehat{p}, E) = \begin{cases} \langle \emptyset, \{1\} \rangle & \text{if } S^1 \in E \\ Can^+(\widehat{p}, E) & \text{if } S^0 \in E \\ \langle \emptyset, \{1\} \rangle \cup Can(\widehat{p}, E) & \text{if } S^\perp \in E. \end{cases}$$

If the second term of a sequence is selected, then the sequence behaves like its selected branch (control cannot reach the first branch without first exiting the sequence):

$$Must(p\,;\,\widehat{q}, E) = Must(\widehat{q}, E)$$
$$Can(p\,;\,\widehat{q}, E) = Can(\widehat{q}, E)$$
$$Can^+(p\,;\,\widehat{q}, E) = Can^+(\widehat{q}, E).$$

When the first branch of the sequence $p\,;\,q$ is selected, we analyze q only if p must (resp. can) terminate, in which case the completion code 0 of p is discarded. Note the complexity of the Can^+ definition, whose cases correspond to cases of both *Must* and *Can*:

$Must(\widehat{p}\,;q,E) =$
$$\begin{cases} Must(\widehat{p},E) & \text{if } 0 \notin Must_K(\widehat{p},E), \\ \left\langle \begin{array}{c} Must_S(\widehat{p},E) \cup Must_S(q,E) \\ Must_K(q,E) \end{array} \right\rangle & \text{otherwise.} \end{cases}$$

$Can(\widehat{p}\,;q,E) =$
$$\begin{cases} Can(\widehat{p},E) & \text{if } 0 \notin Can_K(\widehat{p},E), \\ \left\langle \begin{array}{c} Can_S(\widehat{p},E) \cup Can_S(q,E) \\ Can_K(\widehat{p},E) \setminus \{0\} \cup Can_K(q,E) \end{array} \right\rangle & \text{otherwise.} \end{cases}$$

$Can^+(\widehat{p}\,;q,E) =$
$$\begin{cases} Can^+(\widehat{p},E) & \text{if } 0 \notin Can_K^+(\widehat{p},E), \\ \left\langle \begin{array}{c} Can_S^+(\widehat{p},E) \cup Can_S^+(q,E) \\ Can_K^+(\widehat{p},E) \setminus \{0\} \cup Can_K^+(q,E) \end{array} \right\rangle & \text{if } 0 \in Must_K(\widehat{p},E), \\ \left\langle \begin{array}{c} Can_S^+(\widehat{p},E) \cup Can_S(q,E) \\ Can_K^+(\widehat{p},E) \setminus \{0\} \cup Can_K(q,E) \end{array} \right\rangle & \text{otherwise.} \end{cases}$$

A loop is the infinite sequence of instances of a given statement. Therefore, the rules for `loop` are derived from those of the sequence by taking into account the knowledge the loop body cannot terminate instantaneously:

$Must(\texttt{loop}\ \widehat{p}\ \texttt{end}, E) =$
$$\begin{cases} Must(\widehat{p},E) & \text{if } 0 \notin Must_K(\widehat{p},E), \\ \left\langle \begin{array}{c} Must_S(\widehat{p},E) \cup Must_S(p,E) \\ Must_K(p,E) \end{array} \right\rangle & \text{otherwise.} \end{cases}$$

$Can(\texttt{loop}\ \widehat{p}\ \texttt{end}, E) =$
$$\begin{cases} Can(\widehat{p},E) & \text{if } 0 \notin Can_K(\widehat{p},E), \\ \left\langle \begin{array}{c} Can_S(\widehat{p},E) \cup Can_S(p,E) \\ Can_K(\widehat{p},E) \setminus \{0\} \cup Can_K(p,E) \end{array} \right\rangle & \text{otherwise.} \end{cases}$$

$Can^+(\texttt{loop}\ \widehat{p}\ \texttt{end}, E) =$
$$\begin{cases} Can^+(\widehat{p},E), & \text{if } 0 \notin Can_K^+(\widehat{p},E), \\ \left\langle \begin{array}{c} Can_S^+(\widehat{p},E) \cup Can_S^+(p,E) \\ Can_K^+(\widehat{p},E) \setminus \{0\} \cup Can_K^+(p,E) \end{array} \right\rangle & \text{if } 0 \in Must_K(\widehat{p},E), \\ \left\langle \begin{array}{c} Can_S^+(\widehat{p},E) \cup Can_S(p,E) \\ Can_K^+(\widehat{p},E) \setminus \{0\} \cup Can_K(p,E) \end{array} \right\rangle & \text{otherwise.} \end{cases}$$

When only one of the branches of a parallel statement is selected, its resumption behavior is given by that of the selected branch:

$$
\begin{aligned}
Must(\widehat{p} \parallel q, E) &= Must(\widehat{p}, E) \\
Can(\widehat{p} \parallel q, E) &= Can(\widehat{p}, E) \\
Can^+(\widehat{p} \parallel q, E) &= Can^+(\widehat{p}, E) \\
\\
Must(p \parallel \widehat{q}, E) &= Must(\widehat{q}, E) \\
Can(p \parallel \widehat{q}, E) &= Can(\widehat{q}, E) \\
Can^+(p \parallel \widehat{q}, E) &= Can(\widehat{q}, E).
\end{aligned}
$$

When both branches of the parallel are selected, the signal potential of the parallel statement is the union of the signal potentials of the branches. The completion code potential is computed using the max function, as we explained earlier:

$$
\begin{aligned}
Must(\widehat{p} \parallel \widehat{q}, E) &= \left\langle \begin{array}{c} Must_S(\widehat{p}, E) \cup Must_S(\widehat{q}, E) \\ \max(Must_K(\widehat{p}, E), Must_K(\widehat{q}, E)) \end{array} \right\rangle \\
Can(\widehat{p} \parallel \widehat{q}, E) &= \left\langle \begin{array}{c} Can_S(\widehat{p}, E) \cup Can_S(\widehat{q}, E) \\ \max(Can_K(\widehat{p}, E), Can_K(\widehat{q}, E)) \end{array} \right\rangle \\
Can^+(\widehat{p} \parallel \widehat{q}, E) &= \left\langle \begin{array}{c} Can_S^+(\widehat{p}, E) \cup Can_S^+(\widehat{q}, E) \\ \max(Can_K^+(\widehat{p}, E), Can_K^+(\widehat{q}, E)) \end{array} \right\rangle .
\end{aligned}
$$

The potential functions for selected trap declarations follow the definition of the potentials of the non-selected terms but replace the unselected body with a selected one:

$$
\begin{aligned}
Must(\text{trap } T \text{ in } \widehat{p} \text{ end}, E) &= \langle\, Must_S(\widehat{p}, E)\,,\, \downarrow Must_K(\widehat{p}, E)\,\rangle \\
Can(\text{trap } T \text{ in } \widehat{p} \text{ end}, E) &= \langle\, Can_S(\widehat{p}, E)\,,\, \downarrow Can_K(\widehat{p}, E)\,\rangle \\
Can^+(\text{trap } T \text{ in } \widehat{p} \text{ end}, E) &= \langle\, Can_S^+(\widehat{p}, E)\,,\, \downarrow Can_K^+(\widehat{p}, E)\,\rangle.
\end{aligned}
$$

The same is true for the selected signal declarations:

$Must(\text{signal } S \text{ in } \widehat{p} \text{ end}, E) =$
$$
\begin{cases}
Must(\widehat{p}, E * S^1) \setminus S & \text{if } S^1 \in Must_S(\widehat{p}, E * S^\bot), \\
Must(\widehat{p}, E * S^0) \setminus S & \text{if } S^1 \notin Can_S^+(\widehat{p}, E * S^\bot), \\
Must(\widehat{p}, E * S^\bot) \setminus S & \text{otherwise.}
\end{cases}
$$

$Can(\text{signal } S \text{ in } \widehat{p} \text{ end}, E) =$
$$
\begin{cases}
Can(\widehat{p}, E * S^0) \setminus S & \text{if } S^1 \notin Can_S(\widehat{p}, E * S^\bot), \\
Can_S(\widehat{p}, E * S^\bot) \setminus S & \text{otherwise.}
\end{cases}
$$

$Can^+(\text{signal } S \text{ in } \widehat{p} \text{ end}, E) =$
$$
\begin{cases}
Can^+(\widehat{p}, E * S^1) \setminus S & \text{if } S^1 \in Must_S(p, E * S^\bot) \\
Can^+(\widehat{p}, E * S^0) \setminus S & \text{if } S^1 \notin Can_S^+(\widehat{p}, E * S^\bot) \\
Can_S^+(\widehat{p}, E * S^\bot) \setminus S & \text{otherwise.}
\end{cases}
$$

4.2.2 Elementary Properties

For any statement p in state $\overline{p}$ under event E, the inductive definitions of the potential functions ensure

$$Must(\overline{p}, E) \subseteq Can^+(\overline{p}, E) \subseteq Can(\overline{p}, E).$$

This expresses the intuition that execution is only performed on statements that cannot be invalidated by code pruning. When the execution of a statement does not block (a completion code is produced), the *Must* and Can^+ functions give the same result:

$$Must_K(\overline{p}, E) = \{k\} \Rightarrow Can^+(\overline{p}, E) = Must(\overline{p}, E).$$

The *Must* potential function is monotonic. More precisely, if E_1 and E_2 are events such that $E_1 \subseteq E_2$, then

$$E_1 \subseteq E_2 \Rightarrow Must(\overline{p}, E_1) \subseteq Must(\overline{p}, E_2).$$

This amounts to saying the execution can only progress when additional information is available.

The *Can* and Can^+ functions are also monotonic but contravariant, as more statements are invalidated when more information is available:

$$E_1 \subseteq E_2 \Rightarrow \begin{cases} Can(\overline{p}, E_1) \supseteq Can(\overline{p}, E_2) \\ Can^+(\overline{p}, E_1) \supseteq Can^+(\overline{p}, E_2) \end{cases}.$$

4.3 Semantic Rules

Once a statement has been determined correct and its completion code and output event computed, we use deduction rules to determine the resulting program state and to represent the behavior of the statement under the form of a behavioral transition.

The rules are split into two categories: *s-rules* start executing a fresh statement p; *r-rules* resume execution from an existing state $\widehat{p}$. When two similar rules apply to both a state $\widehat{p}$ and a standard statement p, we group them into a single *sr-rule* acting on a term $\overline{p}$ (e.g., rules (sr-seq1) and (sr-seq2) below).

$$\texttt{nothing} \xrightarrow[E]{\emptyset, 0} \texttt{nothing} \qquad \text{(s-nothing)}$$

$$\frac{k \notin \{0, 1\}}{\texttt{exit } T(k) \xrightarrow[E]{\emptyset, k} \texttt{exit } T(k)} \qquad \text{(s-exit)}$$

$$\texttt{emit } S \xrightarrow[E]{\{S^1\}, 0} \texttt{emit } S \qquad \text{(s-emit)}$$

SEMANTIC RULES

$$\text{pause} \xrightarrow[E]{\emptyset, 1} \widehat{\text{pause}} \qquad \text{(s-pause)}$$

$$\widehat{\text{pause}} \xrightarrow[E]{\emptyset, 0} \text{pause} \qquad \text{(r-pause)}$$

$$\frac{S^1 \in E \qquad p \xrightarrow[E]{E', k} \overline{p}'}{\text{present } S \text{ then } p \text{ else } q \text{ end} \xrightarrow[E]{E', k} \text{present } S \text{ then } \overline{p}' \text{ else } q \text{ end}}$$
(s-present+)

$$\frac{S^0 \in E \qquad q \xrightarrow[E]{E', k} \overline{q}'}{\text{present } S \text{ then } p \text{ else } q \text{ end} \xrightarrow[E]{E', k} \text{present } S \text{ then } p \text{ else } \overline{q}' \text{ end}}$$
(s-present−)

$$\frac{\widehat{p} \xrightarrow[E]{E', k} \overline{p}'}{\text{present } S \text{ then } \widehat{p} \text{ else } q \text{ end} \xrightarrow[E]{E', k} \text{present } S \text{ then } \overline{p}' \text{ else } q \text{ end}}$$
(r-then)

$$\frac{\widehat{q} \xrightarrow[E]{E', k} \overline{q}'}{\text{present } S \text{ then } p \text{ else } \widehat{q} \text{ end} \xrightarrow[E]{E', k} \text{present } S \text{ then } p \text{ else } \overline{q}' \text{ end}} \quad \text{(r-else)}$$

$$\frac{p \xrightarrow[E]{E', k} \overline{p}'}{\text{suspend } S \text{ when } p \xrightarrow[E]{E', k} \text{suspend } S \text{ when } \overline{p}'} \qquad \text{(s-suspend)}$$

$$\frac{S^1 \in E}{\text{suspend } S \text{ when } \widehat{p} \xrightarrow[E]{\emptyset, 1} \text{suspend } S \text{ when } \widehat{p}} \qquad \text{(r-suspend+)}$$

$$\frac{S^0 \in E \qquad \widehat{p} \xrightarrow[E]{E', k} \overline{p}'}{\text{suspend } S \text{ when } \widehat{p} \xrightarrow[E]{E', k} \text{suspend } S \text{ when } \overline{p}'} \qquad \text{(r-suspend−)}$$

$$\frac{\overline{p} \xrightarrow[E]{E',k} \overline{p}' \quad k \neq 0}{\overline{p}\,;q \xrightarrow[E]{E',k} \overline{p}'\,;q} \qquad \text{(sr-seq1)}$$

$$\frac{\overline{p} \xrightarrow[E]{E',0} p \quad q \xrightarrow[E]{F',l} \overline{q}'}{\overline{p}\,;q \xrightarrow[E]{E'\cup F',l} p\,;\overline{q}'} \qquad \text{(sr-seq2)}$$

$$\frac{\widehat{q} \xrightarrow[E]{E',l} \overline{q}'}{p\,;\widehat{q} \xrightarrow[E]{E',l} p\,;\overline{q}'} \qquad \text{(r-seq3)}$$

$$\frac{\overline{p} \xrightarrow[E]{E',k} \overline{p}' \quad k \neq 0}{\text{loop } \overline{p} \text{ end} \xrightarrow[E]{E',k} \text{loop } \overline{p}' \text{ end}} \qquad \text{(sr-loop)}$$

$$\frac{\widehat{p} \xrightarrow[E]{E',0} p \quad p \xrightarrow[E]{E'',k} \overline{p}' \quad k \neq 0}{\text{loop } \widehat{p} \text{ end} \xrightarrow[E]{E'\cup E'',k} \text{loop } \overline{p}' \text{ end}} \qquad \text{(r-do-loop)}$$

$$\frac{p \xrightarrow[E]{E',k} \overline{p}' \quad q \xrightarrow[E]{F',l} \overline{q}'}{p \parallel q \xrightarrow[E]{E'\cup F', \max(k,l)} \overline{p}' \parallel \overline{q}'} \qquad \text{(s-par-both)}$$

$$\frac{\widehat{p} \xrightarrow[E]{E',k} \overline{p}' \quad \widehat{q} \xrightarrow[E]{F',l} \overline{q}'}{\widehat{p} \parallel \widehat{q} \xrightarrow[E]{E'\cup F', \max(k,l)} \overline{p}' \parallel \overline{q}'} \qquad \text{(r-par-both)}$$

$$\frac{\widehat{p} \xrightarrow[E]{E',k} \overline{p}'}{\widehat{p} \parallel q \xrightarrow[E]{E',k} \overline{p}' \parallel q} \qquad \text{(r-par-left)}$$

$$\frac{\widehat{q} \xrightarrow[E]{E',l} \overline{q}'}{p \parallel \widehat{q} \xrightarrow[E]{E',l} p \parallel \overline{q}'} \qquad \text{(r-par-right)}$$

$$\frac{\overline{p} \xrightarrow[E]{E',k} \overline{p}' \quad k=0 \text{ or } k=2}{\text{trap } T \text{ in } \overline{p} \text{ end} \xrightarrow[E]{E',0} \text{trap } T \text{ in } p \text{ end}} \qquad \text{(sr-trap-term)}$$

SEMANTIC RULES 71

$$\frac{\overline{p} \xrightarrow[E]{E',1} \widehat{p}}{\text{trap } T \text{ in } \overline{p} \text{ end} \xrightarrow[E]{E',1} \text{trap } T \text{ in } \widehat{p} \text{ end}} \quad \text{(sr-trap-pause)}$$

$$\frac{\overline{p} \xrightarrow[E]{E',k} \overline{p}' \quad k > 2}{\text{trap } T \text{ in } \overline{p} \text{ end} \xrightarrow[E]{E',\downarrow k} \text{trap } T \text{ in } \overline{p}' \text{ end}} \quad \text{(sr-trap-prop)}$$

$$\frac{S \in \mathit{Must}_S(\overline{p}, E * S^\perp) \quad \overline{p} \xrightarrow[E*S^1]{E',k} \overline{p}'}{\text{signal } S \text{ in } \overline{p} \text{ end} \xrightarrow[E]{E'\backslash S,k} \text{signal } S \text{ in } \overline{p}' \text{ end}} \quad \text{(sr-sig+)}$$

$$\frac{S \notin \mathit{Can}^+_S(\overline{p}, E * S^\perp) \quad \overline{p} \xrightarrow[E*S^0]{E',k} \overline{p}'}{\text{signal } S \text{ in } \overline{p} \text{ end} \xrightarrow[E]{E'\backslash S,k} \text{signal } S \text{ in } \overline{p}' \text{ end}} \quad \text{(sr-sig-)}$$

These rules formalize the intuitive semantics given in Chapter 2:

- The rules concerning the simple statements—(s-nothing), (s-exit), (s-emit), (s-pause), and (r-pause)—are trivial.

- The presence test start rules (s-present+) and (s-present−) give control to a branch based on the presence of the signal. The resume rules (r-then) and (r-else) restart the selected test branch after it paused.

- The suspend start rule (s-suspend) gives control to its body and returns the completion code and output event generated by the body. The suspend resume rule (r-suspend+) describes the case where the suspending signal is present so the body is frozen. The rule (r-suspend−) applies when the signal is absent.

- Rule (sr-seq1) states a sequence pauses if p pauses and that the sequence propagates the traps exited by p. Rule (sr-seq2) states control is instantaneously transferred to q if p terminates. Rule (r-seq3) shows how control is dispatched to the second branch when it is selected.

- When the body of a loop terminates, the rule (r-do-loop) immediately restarts it. In all other cases, the rule (sr-loop) applies, requiring the loop body to not terminate instantaneously.

- The rules (s-par-both) and (r-par-both) apply when both branches of a parallel execute during an instant (i.e., when the parallel is started or when both branches continue running). In both cases, the completion codes of the branches are combined using the max operator. When only one of the branches of a parallel is selected, rules (r-par-left) and (r-par-right) apply.

- Rule (sr-trap-term) applies when the body of a trap declaration terminates or exits the declared trap, in which case the entire construct terminates. When the body of a trap declaration pauses, the trap does too, as expressed by rule (sr-trap-pause). In all other cases, a trap code must be decremented and propagated to the environment, using rule (sr-trap-prop).

- The rule (sr-sig+) establishes a signal present; (sr-sig−) establishes a signal absent. Both rules apply at the level of signal declaration statements: a decision about a signal is only made in the context of the complete scope of a signal—the body of a signal declaration.

4.4 Proof

As in any logic, a constructive behavior *proof* is a sequence of deductions using the rules above. As usual, proofs are represented by stacking inferences. For example,

$$\cfrac{\texttt{emit } S \xrightarrow[E]{\{S^1\},0} \texttt{emit } S \qquad \cfrac{p \xrightarrow[E]{E_1,0} p \quad q \xrightarrow[E]{E_2,k} q}{p \,;\, q \xrightarrow[E]{E_1 \cup E_2, k} p \,;\, q} \qquad \texttt{emit } T \xrightarrow[E]{\{T^1\},0} \texttt{emit } T}{\texttt{emit } S \,;\, \texttt{emit } T \xrightarrow[E]{\{S^1, T^1\},0} \texttt{emit } S \,;\, \texttt{emit } T}$$

As we shall see in the next chapter, the constructive operational semantics offers another way to determine the behavioral transition of a statement.

4.5 Determinism

All the deduction rules and potential function definitions are deterministic. Therefore, the constructive semantics itself is deterministic, meaning that it assigns at most one behavior to any statement or program for given state and input event:

$$\left. \begin{array}{c} \overline{p} \xrightarrow[E]{E',k'} \overline{p}' \\ \overline{p} \xrightarrow[E]{E'',k''} \overline{p}'' \end{array} \right\} \Rightarrow \left\{ \begin{array}{l} E' = E'' \\ k' = k'' \\ \overline{p}' = \overline{p}'' \end{array} \right.$$

Furthermore, there is a unique proof for every transition.

4.6 Loop-Safe Programs. Completion Code Potentials

In rule (r-do-loop), the side condition $k \neq 0$ prevents the body of a loop from starting, terminating, and then starting again during the same instant. Such cases are rejected as incorrect, as they correspond to loops that would have to perform an unbounded number of iterations in an instant.

Esterel compilers generally prohibit instantaneous loops through a simple static restriction. This makes the side condition superfluous. For this, we define the set $K_s(p)$ of *potential start completion codes* of p as follows.

$$
\begin{aligned}
K_s(\texttt{nothing}) &= \{0\} \\
K_s(\texttt{pause}) &= \{1\} \\
K_s(\texttt{exit T}(k)) &= \{k\} \\
K_s(\texttt{emit S}) &= \{0\} \\
K_s(\texttt{present } S \texttt{ then } p \texttt{ else } q \texttt{ end}) &= K_s(p) \cup K_s(q) \\
K_s(\texttt{suspend } S \texttt{ when } p) &= K_s(p) \\
K_s(p \,;\, q) &= \begin{cases} (K_s(p) \setminus \{0\}) \cup K_s(q), \\ \quad \text{if } 0 \in K_s(p) \\ K_s(p), \text{ otherwise} \end{cases} \\
K_s(\texttt{loop } p \texttt{ end}) &= K_s(p) \setminus \{0\} \\
K_s(p \,\|\, q) &= \max(K_s(p), K_s(q)) \\
K_s(\texttt{trap T in } p \texttt{ end}) &= \downarrow K_s(p) \\
K_s(\texttt{signal } S \texttt{ in } p \texttt{ end}) &= K_s(p).
\end{aligned}
$$

We symmetrically define the *potential resumption completion code set* of p, called $K_d(p)$. It is the set of codes that p can return in all instants but the first.

$$
\begin{aligned}
K_d(\texttt{nothing}) &= \emptyset \\
K_d(\texttt{pause}) &= \{0\} \\
K_d(\texttt{exit T}(k)) &= \emptyset \\
K_d(\texttt{emit S}) &= \emptyset \\
K_d(\texttt{present } S \texttt{ then } p \texttt{ else } q \texttt{ end}) &= K_d(p) \cup K_d(q) \\
K_d(\texttt{suspend } S \texttt{ when } p) &= K_d(p) \cup \{1\} \\
K_d(p\,;\,q) &= \begin{cases} K_d(p), \text{ if } 0 \notin K_s(p) \cup K_d(p) \\ K_d(q) \cup K_d(p), \\ \quad \text{if } 0 \in K_s(p) \setminus K_d(p) \\ (K_d(p) \setminus \{0\}) \cup K_s(q) \cup K_d(q), \\ \quad \text{otherwise} \end{cases} \\
K_d(\texttt{loop } p \texttt{ end}) &= K_d(p\,;\,p) \\
K_d(p \parallel q) &= \max(K_d(p), K_d(q)) \\
K_d(\texttt{trap T in } p \texttt{ end}) &= \downarrow K_d(p) \\
K_d(\texttt{signal } S \texttt{ in } p \texttt{ end}) &= K_d(p).
\end{aligned}
$$

Following the inductive definition of the semantics, it is easy to see that $K_s(p)$ is a superset of the set of completion codes a statement can return in its first execution instant, and that $K_d(p)$ is a superset of the set of completion codes a statement can return in instants where it is resumed:

$$
p \xrightarrow[E]{E',k} p' \Rightarrow k \in K_s(p)
$$

$$
\overline{p} \xrightarrow[E]{E',k} \overline{p}' \Rightarrow k \in K_d(p).
$$

We can now define loop-safe programs.

Definition 1 *A program P of body p is* loop-safe *if, for each sub-statement "*`loop q end`*" of p,* $0 \notin K_s(q)$.

In practice, loop-safety is not a very restrictive condition and it makes life easier. However, we know of one case where users find it a little annoying. Assume that two input signals I and J are known to be incompatible, i.e., never present in the same instant. In full Esterel, this is asserted by writing

```
relation I # J;
```

Then, the following loop-unsafe program is correct because the direct path from `loop` to "`end loop`" cannot be taken.

```
loop
    present I else
```

```
        p       % non-instantaneous
      end present;
      present J else
        q       % non-instantaneous
      end present
   end loop
```

A workaround to make any program statically loop-safe is to add a **pause** statement in parallel with the loop body.

```
   loop
      present I else
        p       % non instantaneous
      end present;
      present J else
        q       % non instantaneous
      end present
   ||
      pause
   end loop
```

4.7 Program Behavior

We have now completed the definition of the constructive behavioral semantics for Esterel statements. In this section we explain how the transitions of the body statement of a program translate into transitions of the program itself.

The semantics of a program is most easily defined when no interface signal is both emitted and tested. We start by defining the semantics in this case, and then we explain how the more complex cases can be reduced to the simple one.

Let P be a Kernel Esterel program with body p where no interface signal can be both tested and emitted. Assume that I is the set of input signals of P, and that O is the set of output signals. Let E_I be an event over I (an input event) and E_O an event over O. Then

$$P' \xrightarrow[E_I]{E_O} P'' \Leftrightarrow \exists k \in \{0,1\}, \exists E \text{ such that } \overline{p}' \xrightarrow[E_I]{E,k} \overline{p}'' \text{ and } E/O = E_O$$

where $\overline{p}'$ (resp. $\overline{p}''$) is the state of p in P' (resp. P''), and where E ranges over events over $I \cup O$.

Now, consider when an input signal can be emitted or an output signal can be tested by the program body. In that case, the above definition is not appropriate since it does not generate the needed feedback of the output signals back to the body execution environment. The following two Esterel programs illustrate the problem.

```
module IN_PROBLEM:
input I;
output O;

  emit I ;
  present I then emit O end

end module

module OUT_PROBLEM:
output O1, O2;

  emit O1
||
  present O1 then emit O2 end

end module
```

In the first case, with only the current rules, the signal I would be emitted but the **present** statement would not be informed since the sequence rule (sr-seq2) performs neither signal feedback nor potential computations. Signal O would not be emitted if the environment does not produce the input signal I. In the second example, signal O1 would emitted, but not O2.

To correctly define the semantics of such programs, we need to add feedback for interface signals. We do that by transforming the programs by interface signal duplication and addition of trivial *connection code*, so that no interface signal can be both tested and emitted by the program body. The transformation is straightforward. If the input signal I can be internally emitted by the body statement p, we replace p with

```
signal I_local in
  p′
||
  loop present I then emit I_local ; pause end
end
```

where p' is obtained from p by replacing all occurrences of I by I_local. If the output signal O can be internally tested by p, then p becomes

```
signal O_local in
  p′
||
  loop present O_local then emit O ; pause end
end
```

where all occurrences of I are replaced in p' with I_local.

The previous transformations can be applied only on programs having no **inputoutput** signal or shared variable. In cases where such signals are present, we need to replace each of them with a pair formed of an input signal and an output signal.

5

Constructive Operational Semantics

The previous chapter showed how macrostep transitions can be used to represent the behavior of a statement during an execution instant. A reaction corresponds to one behavioral transition. Such a presentation of the semantics has the advantage of succinctness, but does not define the key notions of *activation*, *execution*, and *causality*, which are essential for efficient code generation. The objective of the constructive operational semantics is to define the three notions while remaining fully compatible with the constructive behavioral semantics. The COS computes a reaction by chaining elementary microsteps whose aggregation computes the behavioral transition. Technically, the COS borrows many components from the CBS and its rules can be viewed as refinements of the behavioral ones.

Handling data is important for compiling actual Esterel programs; the COS includes data. We did not address data in the constructive behavioral semantics because it would be very tedious to handle in that setting.

The chapter is organized as follows. We first define the COS terms, which refine the behavioral terms with a notation for control flow. Then we explain how data is represented in stores. Next, we define the semantic rules and explain how the *Can* potential function is used to determine signal absence and shared variable synchronization. Finally, we explain how microsteps are chained together into full reactions and define the relationship between the Operation Semantics and constructive behavioral semantics.

5.1 Microsteps

The *constructive operational semantics* of Esterel gives the semantics of a statement or program with *microstep transition rules* of the form

$$\dot{p}, data \xrightarrow[E]{e,k} \dot{p}', data'.$$

In these rules, $\dot{p}$ and $\dot{p}'$ are obtained by adding bullets to p at instantaneous control flow positions, which act as program counters. In the COS, transitions perform elementary control flow propagation. Many of them must be chained to compute a complete behavioral transition, as in the following data-less example

$$\bullet(\text{emit } S \text{ ; emit } T \text{ ; pause}) \xrightarrow[E]{\bot,\bot} (\bullet\text{emit } S) \text{ ; emit } T \text{ ; pause}$$

$$\xrightarrow[E]{S,\bot} \text{emit } S \text{ ; } (\bullet\text{emit } T) \text{ ; pause}$$

$$\xrightarrow[E]{T,\bot} \text{emit } S \text{ ; emit } T \text{ ; } \bullet\text{pause}$$

$$\xrightarrow[E]{\bot,1} \text{emit } S \text{ ; emit } T \text{ ; } \widehat{\text{pause}}.$$

The bullet progresses to represent the execution of the statement and disappears when the reaction is complete.

The terms $\dot{p}$ and $\dot{p}'$ not only represent the status of the **pause** statements but also store the intermediate control flow between microsteps within the computation instant. The stores *data* and *data'* define the value of the data variables used in p, before and after the microstep. The input event E defines the signal environment. If the transition represents the completion of statement p for the current execution instant, then k is the integer completion code. Otherwise, $k = \bot$. If p emits the signal S during the transition, then $e = S$. Otherwise, $e = \bot$. In the COS, at most one signal can be emitted during one micro-step.

The COS refines the control and potential structures of the constructive behavioral semantics. The constructive behavioral semantics is organized around the analysis of potentials, which determines statement correctness, signal behavior, and completion codes. In the COS, the focus shifts to microstep rules: the *Must* potential calculation is replaced with actual calculation, and *Can* alone is needed to determine signal absence and shared variable synchronization; the Can^+ potential becomes useless.

5.2 COS Terms

The tags that decorate the semantic terms represent two complementary types of control information: the behavioral state of the Esterel program, defined in the previous chapter (the status of its **pause** statements between reactions); and the microstep execution status, which is the current control flow progress during the execution of a reaction.

The double decoration of the semantic terms is complex, but not without reason. Intuitively, the behavioral status of a statement completely determines the computation of an instant if the statement is executed during that instant. On the other hand, even if our statement retained control from the previous instant, its resumption in the current instant is subject to preemption. The execution of the preemption tests and the actual execution of statements must respect the hierarchy of the Esterel program. This order is enforced using "program counter" decorations, representing control flow progress.

In Chapter 8, we show how the two classes of decorations correspond to components of our GRC intermediate representation for Esterel programs, i.e., the state representation and the control-flow graph.

5.2.1 Control Flow Propagation

The microstep execution status is represented by bullets that act as program counters. The following example shows how a reaction is constructed through control propagation.

$$\bullet(\text{emit } S\,;\,\text{emit } T) \xrightarrow[E]{\bot,\bot} (\bullet\text{emit } S)\,;\,\text{emit } T$$
$$\xrightarrow[E]{S,\bot} \text{emit } S\,;\,\bullet\text{emit } T$$
$$\xrightarrow[E]{T,0} \text{emit } S\,;\,\text{emit } T.$$

The process includes three microsteps. First, the sequence starts by starting its first component. The program counter (the bullet) changes its position. No signal is emitted during the transition, and there is no completion code. Next, "emit S" is executed and signal S is emitted. There is no completion code because the execution of the sequence statement is not completed. The next step completes the execution of the instant with code 0 and generates the signal T.

Compare the previous sequence of microsteps with the deduction of the corresponding macrostep transition in the constructive behavioral semantics, given in Section 4.4. The sequence of microsteps can be seen as an operational way to construct the transition.

5.2.2 State-Dependent Behavior

Recall from Section 4.1.2 that the state of an Esterel program, which is transmitted from one execution instant to the next, is defined by the status of its **pause** statements. A **pause** statement is tagged as $\widehat{\textbf{pause}}$ to indicate that the control flow *paused* there. In this case the **pause** statement is *selected*. The notion of selection is extended naturally to composed Esterel statements:

a statement is selected if it contains a tagged $\widehat{\text{pause}}$ statement. We also say in this case that the *selection status* of the statement is *true*.

The execution of a selected statement always starts with the decoding of the program state. The program counter(s) descend the hierarchy of the syntax tree, from a selected statement to its selected children. The decoding ends at selected $\widehat{\text{pause}}$ statements or when a preemption test succeeds. The following example illustrates the decoding process.

$$\bullet(\widehat{\text{pause}}\,;\,\text{pause}) \xrightarrow[E]{\bot,\bot} (\bullet\widehat{\text{pause}})\,;\,\text{pause}$$

$$\xrightarrow[E]{\bot,\bot} \text{pause}\,;\,\bullet\text{pause}$$

$$\xrightarrow[E]{\bot,1} \text{pause}\,;\,\widehat{\text{pause}}$$

The program counter first decodes the state of the sequence statement and descends to the selected $\widehat{\text{pause}}$ statement. Then, it resumes that $\widehat{\text{pause}}$ statement, which terminates and passes control to the second (unselected) pause statement. This one pauses, i.e., completes with code 1 and receives a tag, so that the sequence pauses. The microstep sequence is completed since there is no bullet left to propagate. The corresponding macrostep transition in the CBS is a collapse of these three microsteps.

$$\widehat{\text{pause}}\,;\,\text{pause} \xrightarrow[E]{\emptyset,1} \text{pause}\,;\,\widehat{\text{pause}}$$

The next example illustrates preemption. Here, the state of the sequence statement is not decoded; the presence of signal S preempts the process.

$$\bullet\text{suspend pause}\,;\,\widehat{\text{pause}} \text{ when } S \xrightarrow[\{S\}]{\bot,1} \text{suspend pause}\,;\,\widehat{\text{pause}} \text{ when } S$$

The corresponding transition in the reference semantics is

$$\text{suspend pause}\,;\,\widehat{\text{pause}} \text{ when } S \xrightarrow[\{S\}]{\emptyset,1} \text{suspend pause}\,;\,\widehat{\text{pause}} \text{ when } S.$$

5.2.3 Syntax of Semantic Terms

The terms of the COS extend the behavioral terms defined in Section 4.1.2. With the same definition, the term $\widehat{p}$ represents a selected term over the statement p, while $\overline{p}$ represents a term over p that may or may not be selected.

The term $\dot{p}$ represents a term over p containing a "program counter" (bullet). We write $\ddot{p}$ to indicate that we do not know whether p contains a program counter. Semantic terms are obtained by combining the two notations, so we classify the semantic terms over p into nine classes, listed in the following table.

Contains	Is selected		
bullets	no	yes	maybe
no	p	$\hat{p}$	$\overline{p}$
yes	$\dot{p}$	$\dot{\hat{p}}$	$\dot{\overline{p}}$
maybe	$\ddot{p}$	$\ddot{\hat{p}}$	$\ddot{\overline{p}}$

For example, $\dot{\hat{p}}$ represents a selected term over p that contains one or more bullets. When confusion is possible in a semantic rule or in a longer derivation, we add indices to indicate different terms of the same kind.

We defined the behavioral terms $\hat{p}$ and $\overline{p}$ in Section 4.1.2, so here we only give the syntax of the control and combined terms. In addition to "program counter" bullets, these terms may also contain decorations that represent the current status of signals, parallel statements, and loops.

$$
\begin{aligned}
\dot{p} \ ::=\ & \bullet \overline{p} \\
| \ & \dot{p}\,;\,p \\
| \ & p\,;\,\dot{p} \\
| \ & \dot{p}\ {}^\perp\|{}^\perp\ \dot{p} \\
| \ & \dot{p}\ {}^\perp\|{}^l\ \overline{p} \\
| \ & \overline{p}\ {}^k\|{}^\perp\ \dot{p} \\
| \ & \overline{p}\ {}^k\|{}^l\ \overline{p} \\
| \ & \texttt{loop}^{lstat}\ \dot{p}\ \texttt{end} \\
| \ & \texttt{suspend}\ \dot{p}\ \texttt{when}\ S \\
| \ & \texttt{present}\ S\ \texttt{then}\ \dot{p}\ \texttt{else}\ p\ \texttt{end} \\
| \ & \texttt{present}\ S\ \texttt{then}\ p\ \texttt{else}\ \dot{p}\ \texttt{end} \\
| \ & \texttt{trap}\ T\ \texttt{in}\ \dot{p}\ \texttt{end} \\
| \ & \texttt{var}\ v\ \texttt{in}\ \dot{p}\ \texttt{end} \\
| \ & \texttt{shared}\ s\ \texttt{in}\ \dot{p}\ \texttt{end} \\
| \ & \texttt{signal}\ S^{sstat}\ \texttt{in}\ \dot{p}\ \texttt{end} \\
\ddot{p} \ ::=\ & \dot{p} \\
| \ & \overline{p}
\end{aligned}
$$

The decorations on the || operator give the current completion status of each branch: $\perp$ means that the branch did not yet complete its computation for the current reaction. An integer code denoted by k or l in these rules represents a branch that completed with the given code. The term $\overline{q}\ {}^k\|{}^l\ \overline{r}$ represents an uncompleted parallel whose branches have completed. A parallel in this state requires one more microstep to synchronize the completion codes of its branches and produce its final completion code $\max(k,l)$.

The status $lstat$ associated with each loop is used to semantically prohibit the instantaneous execution of the loop's body and thus avoid diverging computation. The status can be $stop$ or go; $stop$ indicates the body must not be restarted in the current reaction. When the loop statement in started in an instant, its status is set to *stop*. If it is resumed, then its status is set to *go* since the loop body can be restarted once. Note that this mechanism is not needed in the constructive behavioral semantics, where transitions cover entire execution instants. Reasoning about individual microsteps exposes the problem of instantaneous loop body termination, which any implementation needs to address.

The status *sstat* of a signal can be $\bot$ (not yet defined), 1 (present), or 0 (absent). The status is set to $\bot$ when the `signal` statement is started or resumed. It becomes 1 if it is emitted, and 0 if all emissions have been ruled out by the control flow.

5.3 Data Representation

The ability to handle data and its non-constructive causality is a fundamental advantage of the COS. Whenever we define the semantics of a statement that manipulates data, we represent the status of the variables using stores. The following transition shows how an assignment changes the value of a non-shared variable v.

$$\bullet v := expr, data \xrightarrow[E]{\bot, 0} v := expr, data[v \leftarrow [\![expr]\!]]$$

Here, $[\![expr]\!]$ is the value obtained by evaluating the expression *expr* in the data context *data*. Note how the flow of control determines a change in the value stored for v. The value, of type $type_v$, associated by the store *data* to the non-shared variable v can be read with $data(v)$.

Things are more complicated for *shared variables*, where we also encode the needed synchronization information on the store. For every Esterel shared variable s, we use two variables on the store. The store variable s_{value} gives the value of s, and the store variable s_{status} gives its synchronization status, telling us whether s_{value} can be read or updated. The possible values of s_{status} are

ready, meaning all write actions on s have either been executed or invalidated by control flow. Thus, we can read its value with $data(s_{value})$.

old, meaning no write action has been yet executed, so the variable still has the value of the former reaction. It cannot be read because write actions may still be executed.

new, meaning the variable has already been modified by write actions, but it cannot yet be read because other actions may still modify it.

Shared variables and their synchronization also require changing the potential function *Can*. In addition to the two components (Can_S and Can_K) used in the Constructive Behavioral Semantics, we need a third one: Can_V.

$$Can(\dot{p}, E) = < Can_S(\dot{p}, E), Can_K(\dot{p}, E), Can_V(\dot{p}, E) >$$

The new component determines which shared variable assignments are not yet invalidated by the current state and input event. When a shared variable cannot be assigned again within a reaction, its status can be changed to *ready*, allowing it to be read. We define the Can_V function in Section 5.5.

5.4 Semantic Rules

With sixty transition rules, the COS is far larger than the constructive behavioral semantics. This is because behaviors are divided into basic computations so all common control flow operations (start, completion, etc.) are represented separately for each statement. All the rules are necessary to ensure the completeness of the semantics.

To facilitate the understanding of the semantics and emphasize the relation with the constructive behavioral semantics, we divide the rules in two sets: those for Pure Esterel and those that handle data.

Statements in the first set do not change or read the stores (sub-statements can, however). Furthermore, removing all reference to data rules results in a dataless constructive operational semantics for Pure Esterel.

5.4.1 Rules for Pure Esterel Primitives

Rule for the no-operation statement

A no-operation statement **nothing** simply relinquishes control. No variables are changed.

$$\bullet\texttt{nothing}, data \xrightarrow[E]{\bot,0} \texttt{nothing}, data \qquad (1)$$

Pause rules

Reaching a **pause** pauses execution. Variables are unchanged.

$$\bullet\texttt{pause}, data \xrightarrow[E]{\bot,1} \widehat{\texttt{pause}}, data \qquad (2)$$

When resumed, a selected $\widehat{\texttt{pause}}$ statement behaves like a **nothing** statement. It terminates instantly. The selection mark is deleted.

$$\bullet\widehat{\texttt{pause}}, data \xrightarrow[E]{\bot,0} \texttt{pause}, data \qquad (3)$$

Sequence rules

Starting a sequence consists of starting its first branch. This does not modify variables.

$$\bullet(p \,;\, q), data \xrightarrow[E]{\bot,\bot} (\bullet p) \,;\, q, data$$

Similarly, resuming a sequence when the first branch is selected, gives control to the first branch and does not modify data.

$$\bullet(\widehat{p}\ ;\ q), data \xrightarrow[E]{\perp,\perp} (\bullet\widehat{p})\ ;\ q, data$$

These two transition rules can be merged:

$$\bullet(\overline{p}\ ;\ q), data \xrightarrow[E]{\perp,\perp} (\bullet\overline{p})\ ;\ q, data \qquad (4)$$

We will often use this trick to reduce the number of rules.

Internal transitions of a sub-statement are transformed into transitions of the composite statement itself. The transformation can be "passive," meaning the composed statement merely informs the environment about the evolution of its child. The following rule shows how internal transitions of the first branch of a sequence become transitions of the sequence itself.

$$\frac{\dot{p}, data \xrightarrow[E]{e,k} \ddot{p}, data'\ \ k \neq 0}{\dot{p}\ ;\ q, data \xrightarrow[E]{e,k} \ddot{p}\ ;\ q, data'} \qquad (5)$$

Every compound statement has a similar rule.

More interesting are the transitions where computation is performed by the statement. The following rule states that control is given to the second branch of the sequence when the first terminates.

$$\frac{\dot{p}, data \xrightarrow[E]{e,0} p, data'}{\dot{p}\ ;\ q, data \xrightarrow[E]{e,\perp} p\ ;\ (\bullet q), data'} \qquad (6)$$

This involves not only a transition of the first branch, but also control passing through the sequence.

The resume rule applies when the second branch is selected.

$$\bullet(p\ ;\ \widehat{q}), data \xrightarrow[E]{\perp,\perp} p\ ;\ \bullet\widehat{q}, data \qquad (7)$$

The last sequence rule performs microsteps of the second branch.

$$\frac{\dot{q}, data \xrightarrow[E]{e,k} \ddot{q}, data'}{p\ ;\ \dot{q}, data \xrightarrow[E]{e,k} p\ ;\ \ddot{q}, data'} \qquad (8)$$

Parallel rules

Starting a parallel starts its branches and initializes the status of each branch to $\perp$ (not completed). The variables are not modified.

$$\bullet(p\ ||\ q), data \xrightarrow[E]{\perp,\perp} (\bullet p)\ {}^{\perp}||^{\perp}\ (\bullet q), data \qquad (9)$$

Resuming a parallel statement requires three rules, corresponding to both branches being selected, or only one (left or right). When both branches are selected, control is distributed to both. The synchronization decorations on the parallel are set to $\bot$ for both branches, signaling that the completion of both branches is needed to complete the instant.

$$\bullet(\widehat{p} \parallel \widehat{q}), data \xrightarrow[E]{\bot,\bot} (\bullet\widehat{p})^{\bot}\parallel^{\bot} (\bullet\widehat{q}), data \qquad (10)$$

When only the left branch is selected, only it receives control. The synchronization decoration of the right branch is set to 0, as for a terminated branch. Thus, the synchronization rule (13) gives the correct result when the execution of the left branch is completed.

$$\bullet(\widehat{p} \parallel q), data \xrightarrow[E]{\bot,\bot} (\bullet\widehat{p})^{\bot}\parallel^{0} q, data \qquad (11)$$

Similarly, for the case where the right branch is selected,

$$\bullet(p \parallel \widehat{q}), data \xrightarrow[E]{\bot,\bot} p^{\,0}\parallel^{\bot} (\bullet\widehat{q}), data \qquad (12)$$

When the execution of the active parallel branches completes (i.e., when all the synchronization decorations are different from $\bot$), the parallel itself completes.

$$\overline{p}^{\,k}\parallel^{l} \overline{q}, data \xrightarrow[E]{\bot,\max(k,l)} \overline{p} \parallel \overline{q}, data \qquad (13)$$

Microsteps of the branches are interleaved arbitrarily to form executions of the parallel statement. The rule that transforms microsteps of the left branch into microsteps of the parallel statement is

$$\frac{\dot{p}, data \xrightarrow[E]{e,k} \ddot{p}, data'}{\dot{p}^{\,\bot}\parallel^{m} \ddot{q}, data \xrightarrow[E]{e,\bot} \ddot{p}^{\,k}\parallel^{m} \ddot{q}, data'} \qquad (14)$$

For the right branch,

$$\frac{\dot{q}, data \xrightarrow[E]{e,k} \ddot{q}, data'}{\ddot{p}^{\,m}\parallel^{\bot} \dot{q}, data \xrightarrow[E]{e,\bot} \ddot{p}^{\,m}\parallel^{k} \ddot{q}, data'} \qquad (15)$$

Although the interleaving is arbitrary, the result is deterministic.

Loop rules

When a loop is started, its status is initialized to *stop* to forbid the instantaneous termination of the body and therefore its instantaneous restart.

$$\bullet\texttt{loop}\ p\ \texttt{end}, data \xrightarrow[E]{\bot,\bot} \texttt{loop}^{stop}\ \bullet p\ \texttt{end}, data \qquad (16)$$

If the body terminates while the loop status is *stop*, execution is blocked and the program is incorrect. No rewriting rules handle this configuration.

The body of a loop is allowed to terminate in instants where the loop is resumed. The status is initialized to *go*.

$$\bullet \texttt{loop } \widehat{p} \texttt{ end}, data \xrightarrow[E]{\bot,\bot} \texttt{loop}^{go} \bullet \widehat{p} \texttt{ end}, data \qquad (17)$$

In such cases, the loop body is instantaneously restarted upon termination. The loop status is set to *stop* to prohibit a second instantaneous termination of the body.

$$\frac{\dot{p}, data \xrightarrow[E]{e,0} p, data'}{\texttt{loop}^{go} \: \dot{p} \texttt{ end}, data \xrightarrow[E]{e,\bot} \texttt{loop}^{stop} \bullet p \texttt{ end}, data'} \qquad (18)$$

A loop statement completes its execution when its body pauses or exits an trap.

$$\frac{\dot{p}, data \xrightarrow[E]{e,k} \overline{p}, data' \quad k \notin \{\bot, 0\}}{\texttt{loop}^m \: \dot{p} \texttt{ end}, data \xrightarrow[E]{e,k} \texttt{loop } \overline{p} \texttt{ end}, data'} \qquad (19)$$

The loop body performs microsteps as follows:

$$\frac{\dot{p} \xrightarrow[E]{e,\bot} (\dot{p})', data}{\texttt{loop}^m \: \dot{p} \texttt{ end} \xrightarrow[E]{e,\bot} \texttt{loop}^m \: (\dot{p})' \texttt{ end}, data'} \qquad (20)$$

Trap-exit rules

Starting or resuming a trap statement gives control to its body.

$$\bullet \texttt{trap } T \texttt{ in } \overline{p} \texttt{ end}, data \xrightarrow[E]{\bot,\bot} \texttt{trap } T \texttt{ in } \bullet \overline{p} \texttt{ end}, data \qquad (21)$$

Internal microsteps of the body become microsteps of the trap statement.

$$\frac{\dot{p}, data \xrightarrow[E]{e,\bot} (\dot{p})', data'}{\texttt{trap } T \texttt{ in } \dot{p} \texttt{ end}, data \xrightarrow[E]{e,\bot} \texttt{trap } T \texttt{ in } (\dot{p})' \texttt{ end}, data'} \qquad (22)$$

A trap instruction handles completion code 2 by preempting its body. This corresponds to the locally-handled trap exiting (Section 3.3.4). Here, all the decorations of the semantic term are removed to represent the preemption. It is important to note that the trap operator is the only one that globally modifies the selection status of a term.

$$\frac{\dot{\overline{p}}, data \xrightarrow[E]{\perp,2} \overline{p}, data'}{\text{trap } T \text{ in } \dot{\overline{p}} \text{ end}, data \xrightarrow[E]{\perp,0} \text{trap } T \text{ in } p \text{ end}, data'} \tag{23}$$

When the body of the trap terminates, the resulting term is the same.

$$\frac{\dot{\overline{p}}, data \xrightarrow[E]{e,0} p, data'}{\text{trap } T \text{ in } \dot{\overline{p}} \text{ end}, data \xrightarrow[E]{e,0} \text{trap } T \text{ in } p \text{ end}, data'} \tag{24}$$

The trap statement pauses when its body pauses.

$$\frac{\dot{\overline{p}}, data \xrightarrow[E]{\perp,1} \widehat{p}, data}{\text{trap } T \text{ in } \dot{\overline{p}} \text{ end}, data \xrightarrow[E]{\perp,1} \text{trap } T \text{ in } \widehat{p} \text{ end}, data} \tag{25}$$

When the trap body exits a trap that cannot be handled locally (code $k \geq 3$), the trap statement updates the code and passes it to its environment.

$$\frac{\dot{\overline{p}}, data \xrightarrow[E]{\perp,k} \overline{p}, data \quad k \geq 3}{\text{trap } T \text{ in } \dot{\overline{p}} \text{ end}, data \xrightarrow[E]{\perp,\downarrow k} \text{trap } T \text{ in } \overline{p} \text{ end}, data} \tag{26}$$

The definition of the function $\downarrow$ appears in Section 3.3.4.

Raising a trap produces the given completion code $k \notin \{0, 1\}$.

$$\bullet \text{exit } T(k), data \xrightarrow[E]{\perp,k} \text{exit } T(k), data \tag{27}$$

Signal scope rules

Starting or resuming a signal declaration statement gives control to its body and sets the status of the declared signal to $\perp$ (undefined).

$$\bullet \text{signal } S \text{ in } \overline{p} \text{ end}, data \xrightarrow[E]{\perp,\perp} \text{signal } S^{\perp} \text{ in } \bullet\overline{p} \text{ end}, data \tag{28}$$

The signal S becomes present when the body of the signal statement first emits it without instantly completing execution. Later emissions of S have no further effect.

$$\frac{\dot{\overline{p}}, data \xrightarrow[E*S^m]{S,\perp} (\dot{\overline{p}})', data \quad m \in \{\perp, 1\}}{\text{signal } S^m \text{ in } \dot{\overline{p}} \text{ end}, data \xrightarrow[E]{\perp,\perp} \text{signal } S^1 \text{ in } (\dot{\overline{p}})' \text{ end}, data} \tag{29}$$

The signal status is set to absent if it has not yet been emitted and all its emissions have been ruled out by *Can* potential analysis performed on the current control state.

$$\frac{S \notin Can_S(\dot{p}, E * S^\perp)}{\text{signal } S^\perp \text{ in } \dot{p} \text{ end}, data \xrightarrow[E]{\perp,\perp} \text{signal } S^0 \text{ in } \dot{p} \text{ end}, data} \quad (30)$$

The *potential function Can$_S$* is an extension of the potential function of the same name in the constructive behavioral semantics. Similarly, it decides, based on the current microstep state and input event, which signals can be emitted by p. We give the full definition of Can_S in Section 5.5.

Non-completing microsteps of the body that do not generate S can generate either $e = \perp$ or $e = T$ for some other signal T. This event e is returned by the signal declaration.

$$\frac{\dot{p}, data \xrightarrow[E*S^m]{e,\perp} (\dot{p})', data' \quad e \neq S}{\text{signal } S^m \text{ in } \dot{p} \text{ end}, data \xrightarrow[E]{e,\perp} \text{signal } S^m \text{ in } (\dot{p})' \text{ end}, data'} \quad (31)$$

The following rule handles statement completion when no event is generated or the generated event must be passed to the environment. Control leaves the statement.

$$\frac{\dot{p}, data \xrightarrow[E*S^m]{e,k} \overline{p}, data' \quad k \neq \perp, e \neq S}{\text{signal } S^m \text{ in } \dot{p} \text{ end}, data \xrightarrow[E]{e,k} \text{signal } S \text{ in } \overline{p} \text{ end}, data'} \quad (32)$$

The next rule handles statement completion when the internal emission of S is hidden, because no further microstep can read it.

$$\frac{\dot{p}, data \xrightarrow[E*S^m]{S,k} \overline{p}, data' \quad k \neq \perp}{\text{signal } S^m \text{ in } \dot{p} \text{ end}, data \xrightarrow[E]{\perp,k} \text{signal } S \text{ in } \overline{p} \text{ end}, data'} \quad (33)$$

Signal emission

The signal emission rule produces both a signal event and a completion code. Note that in the COSif a transition rule produces both a signal event and a completion code, then the code is 0 and it comes from a signal emission sub-statement.

$$\bullet \text{emit } S, data \xrightarrow[E]{S,0} \text{emit } S, data \quad (34)$$

Suspension rules

Two statements make control flow decisions based on signal status: **suspend** and **present**. The suspension test preempts the resumption of a statement if the guard signal is present.

$$\frac{S^1 \in E}{\bullet\texttt{suspend}\ \widehat{p}\ \texttt{when}\ S, data \xrightarrow[E]{\bot,1} \texttt{suspend}\ \widehat{p}\ \texttt{when}\ S, data} \quad (35)$$

When the signal is absent, control is given to the body, which is resumed.

$$\frac{S^0 \in E}{\bullet\texttt{suspend}\ \widehat{p}\ \texttt{when}\ S, data \xrightarrow[E]{\bot,\bot} \texttt{suspend}\ \bullet\widehat{p}\ \texttt{when}\ S, data} \quad (36)$$

Signal tests block while the status of the signal is $\bot$ because no semantic rule covers that case.

When started, **suspend** gives control to its body.

$$\bullet\texttt{suspend}\ p\ \texttt{when}\ S, data \xrightarrow[E]{\bot,\bot} \texttt{suspend}\ \bullet p\ \texttt{when}\ S, data \quad (37)$$

Once the body takes control, the entire **suspend** statement behaves like its body, including transitions where the body completes.

$$\frac{\dot{p}, data \xrightarrow[E]{e,k} \ddot{p}, data'}{\texttt{suspend}\ \dot{p}\ \texttt{when}\ S, data \xrightarrow[E]{e,k} \texttt{suspend}\ \ddot{p}\ \texttt{when}\ S, data'} \quad (38)$$

Signal test rules

When started, the signal presence test starts one of its branches depending on the signal status. If the signal is present, the "**then**" branch is started.

$$\frac{S^1 \in E}{\bullet\texttt{present}\ S\ \texttt{then}\ p\ \texttt{else}\ q\ \texttt{end}, data \xrightarrow[E]{\bot,\bot} \texttt{present}\ S\ \texttt{then}\ \bullet p\ \texttt{else}\ q\ \texttt{end}, data} \quad (39)$$

When the signal is absent, control is given to the "**else**" branch.

$$\frac{S^0 \in E}{\bullet\texttt{present}\ S\ \texttt{then}\ p\ \texttt{else}\ q\ \texttt{end}, data \xrightarrow[E]{\bot,\bot} \texttt{present}\ S\ \texttt{then}\ p\ \texttt{else}\ \bullet q\ \texttt{end}, data} \quad (40)$$

Control blocks while the status of the signal is $\bot$ since no rule applies.
When resumed, the statement resumes its selected branch.

$$\bullet\texttt{present}\ S\ \texttt{then}\ \widehat{p}\ \texttt{else}\ q\ \texttt{end}, data \xrightarrow[E]{\bot,\bot} \texttt{present}\ S\ \texttt{then}\ \bullet\widehat{p}\ \texttt{else}\ q\ \texttt{end}, data \quad (41)$$

$$\bullet\text{present } S \text{ then } p \text{ else } \widehat{q} \text{ end}, data \xrightarrow[E]{\bot,\bot} \text{present } S \text{ then } p \text{ else } \bullet\widehat{q} \text{ end}, data \tag{42}$$

Once one of the branches take control, the entire **present** statement behaves like that branch, including in transitions where the branch completes. Here are the microsteps of the "**then**" branch:

$$\frac{\dot{p}, data \xrightarrow[E]{e,k} \ddot{p}, data'}{\text{present } S \text{ then } \dot{p} \text{ else } q \text{ end}, data \xrightarrow[E]{e,k} \text{present } S \text{ then } \ddot{p} \text{ else } q \text{ end}, data'} \tag{43}$$

And here are the microsteps in the "**else**" branch:

$$\frac{\dot{q}, data \xrightarrow[E]{e,k} \ddot{q}, data'}{\text{present } S \text{ then } p \text{ else } \dot{q} \text{ end}, data \xrightarrow[E]{e,k} \text{present } S \text{ then } p \text{ else } \ddot{q} \text{ end}, data'} \tag{44}$$

5.4.2 Rules for Data-Handling Primitives

Shared variable scope rules

Starting the statement declaring the shared variable s evaluates $init_s$, initializes s_{value} with the result, sets s_{status} to *old*, and gives control to the body. If $init_s$ is defined by the function call $f(v^1, \ldots, v^n, s^1, \ldots, s^m)$, the rule is

$$\frac{\forall 1 \leq i \leq m, data(s^i_{status}) = ready}{\bullet\text{shared } s \text{ in } p \text{ end}, data \xrightarrow[E]{\bot,\bot} \text{shared } s \text{ in } \bullet p \text{ end}, data'}, \tag{45}$$

where the new status $data'$ of the store is

$$data \begin{bmatrix} s_{status} \leftarrow old \\ s_{value} \leftarrow f(data(v^1), \ldots, data(v^n), data(s^1_{value}), \ldots, data(s^m_{value})) \end{bmatrix}.$$

Resuming a shared variable declaration does not modify the variable value, but changes its status to *old*.

$$\bullet\text{shared } s \text{ in } \widehat{p} \text{ end}, data \xrightarrow[E]{\bot,\bot} \text{shared } s \text{ in } \bullet\widehat{p} \text{ end}, data[s_{status} \leftarrow old] \tag{46}$$

To decide that a shared variable can be read, we test whether write actions can affect it. If this is not possible, the status of the variable is changed to *ready*. This decision is based on the result of the potential function Can_V.

$$\frac{s \notin Can_V(\dot{p}, E) \quad data(s_{status}) \in \{old, new\}}{\text{shared } s \text{ in } \dot{p} \text{ end}, data \xrightarrow[E]{\bot,\bot} \text{shared } s \text{ in } \dot{p} \text{ end}, data[s_{status} \leftarrow ready]} \tag{47}$$

SEMANTIC RULES

Internal microsteps of the body are microsteps of the statement.

$$\frac{\dot{p}, data \xrightarrow{e,k}_{E} \ddot{p}, data'}{\text{shared } s \text{ in } \dot{p} \text{ end}, data \xrightarrow{e,k}_{E} \text{shared } s \text{ in } \ddot{p} \text{ end}, data'} \quad (48)$$

Shared variable emission rules

When the first write action is performed on a shared variable in a given execution instant, its status changes from *old* to *new* and its new value is the result of the function call. The function computation (and the execution of the microstep) can only occur after all the shared variable arguments are ready to be read.

$$\frac{data(s_{status}) = old \quad \forall 1 \leq i \leq m, data(s^i_{status}) = ready}{\bullet s <= f(v^1, \ldots, v^n, s^1, \ldots, s^m), data \xrightarrow{\bot,0}_{E} s <= f(v^1, \ldots, v^n, s^1, \ldots, s^m), data'} \quad (49)$$

where the new status $data'$ of the store is

$$data \left[\begin{array}{l} s_{status} \leftarrow new \\ s_{value} \leftarrow f(data(v^1), \ldots, data(v^n), data(s^1_{value}), \ldots, data(s^m_{value})) \end{array} \right].$$

On subsequent write actions, the status is preserved and the value is updated using the combine function c_s.

$$\frac{data(s_{status}) = new \quad \forall 1 \leq i \leq m, data(s^i_{status}) = ready}{\bullet s <= f(v^1, \ldots, v^n, s^1, \ldots, s^m), data \xrightarrow{\bot,0}_{E} s <= f(v^1, \ldots, v^n, s^1, \ldots, s^m), data'} \quad (50)$$

where $data'$ is obtained from $data$ by assigning to s_{value} the value

$$c_s(s_{value}, f(data(v^1), \ldots, data(v^n), data(s^1_{value}), \ldots, data(s^m_{value}))).$$

Sequential variable rules

Starting the statement declaring the non-shared variable v evaluates $init_v$, initializes v with the result, and gives control to the body. If $init_v$ is defined by the function call $f(v^1, \ldots, v^n, s^1, \ldots, s^m)$, then the rule is

$$\frac{\forall 1 \leq i \leq m, data(s^i_{status}) = ready}{\bullet \text{var } v \text{ in } p \text{ end}, data \xrightarrow{\bot,\bot}_{E} \text{var } v \text{ in } \bullet p \text{ end}, data'}, \quad (51)$$

where $data'$ is obtained from $data$ by assigning to v the value

$$f(data(v^1), \ldots, data(v^n), data(s^1_{value}), \ldots, data(s^m_{value})).$$

Resuming the statement only gives control to the body.

$$\bullet\text{var } v \text{ in } \widehat{p} \text{ end}, data \xrightarrow[E]{\bot,\bot} \text{var } v \text{ in } \bullet\widehat{p} \text{ end}, data \qquad (52)$$

Once started, the sequential variable declaration behaves as its body.

$$\frac{\dot{p}, data \xrightarrow[E]{e,k} \ddot{p}, data'}{\text{var } v \text{ in } \dot{p} \text{ end}, data \xrightarrow[E]{e,k} \text{var } v \text{ in } \ddot{p} \text{ end}, data'} \qquad (53)$$

Here is the variable test rule, when the value of the test variable is *true*. When the "then" branch is started,

$$\frac{data(v) = true}{\bullet\text{if } v \text{ then } p \text{ else } q \text{ end}, data \xrightarrow[E]{\bot,\bot} \text{if } v \text{ then } \bullet p \text{ else } q \text{ end}, data}. \qquad (54)$$

When the variable is *false*, the "else" branch is started.

$$\frac{data(v) = false}{\bullet\text{if } v \text{ then } p \text{ else } q \text{ end}, data \xrightarrow[E]{\bot,\bot} \text{if } v \text{ then } p \text{ else } \bullet q \text{ end}, data} \qquad (55)$$

When resumed, a test resumes its selected then or else branch.

$$\bullet\text{if } v \text{ then } \widehat{p} \text{ else } q \text{ end}, data \xrightarrow[E]{\bot,\bot} \text{if } v \text{ then } \bullet\widehat{p} \text{ else } q \text{ end}, data \qquad (56)$$

$$\bullet\text{if } v \text{ then } p \text{ else } \widehat{q} \text{ end}, data \xrightarrow[E]{\bot,\bot} \text{if } v \text{ then } p \text{ else } \bullet\widehat{q} \text{ end}, data \qquad (57)$$

A test propagates internal transitions of its branches.

$$\frac{\dot{p}, data \xrightarrow[E]{e,k} \ddot{p}, data'}{\text{if } v \text{ then } \dot{p} \text{ else } q \text{ end}, data \xrightarrow[E]{e,k} \text{if } v \text{ then } \ddot{p} \text{ else } q \text{ end}, data'} \qquad (58)$$

$$\frac{\dot{q}, data \xrightarrow[E]{e,k} \ddot{q}, data'}{\text{if } v \text{ then } p \text{ else } \dot{q} \text{ end}, data \xrightarrow[E]{e,k} \text{if } v \text{ then } p \text{ else } \ddot{q} \text{ end}, data'} \qquad (59)$$

The non-shared variable assignment updates the store with the new value of the variable. The transition can only occur after all its shared variable arguments are ready.

$$\frac{\forall 1 \leq i \leq m, data(s^i_{status}) = ready}{\bullet v := f(v^1, \ldots, v^n, s^1, \ldots, s^m), data \xrightarrow[E]{\bot,0} v := f(v^1, \ldots, v^n, s^1, \ldots, s^m), data'}, \qquad (60)$$

where the new status of the *data* variables is defined by

$$data' = data[v \leftarrow f(data(v^1), \ldots, data(v^n), data(s^1_{value}), \ldots, data(s^m_{value}))].$$

5.5 Analysis of Potentials

We saw in the previous section how incremental execution is explicitly represented in the COSas opposed to the use of the *Must* potential function in the constructive behavioral semantics. Therefore, the scope and complexity of the analysis of potentials is reduced here to code pruning, which is needed to decide signal absence and the end of the computation of shared variables.

To represent code pruning, the COS preserves the corresponding *Can* potential function of the constructive behavioral semantics and extends it to cover partially-executed terms, which contain "program counters" and data-handling statements. The form of the *Can* potential function is

$$Can(\dot{p}, E) = < Can_S(\dot{p}, E), Can_K(\dot{p}, E), Can_V(\dot{p}, E) >,$$

where

- $Can_S(\dot{p}, E)$ tells us which signals can be emitted by p starting in its current state;

- $Can_K(\dot{p}, E)$ determines the completion codes p can generate; and

- $Can_V(\dot{p}, E)$ determines which shared variables have assignment statements that are neither executed nor invalidated by the control flow.

The first two components of *Can* play the same role as their constructive behavioral semantics counterparts. The third component is an essential part of the data access synchronization protocol. A shared variable can be read only when all its assignments have been executed or invalidated. Hence, the status of a shared variable s can be set to *ready*, using rule (47), only when $s \notin Can_V(\dot{p}, E)$.

The *Can* function extends the constructive behavioral semantics function of the same name, meaning that for statements containing no control (terms of the form $\overline{p}$), the components Can_S and Can_K are given by the rules in Section 4.2. When computing the potential of partially-executed terms of the form $\dot{p}$, certain statements may need to be pruned due to the current control state. To take this into account, we compute the potential of such terms in two phases: the the not-yet-executed part of the statement is isolated (the computation is reduced to non-dotted terms); then the potential of non-dotted terms is computed using the rules borrowed from the reference semantics and their data-handling extensions, defined below.

Notation

The inclusion predicate $\subseteq$ and the union operator $\cup$ are extended component-wise on triplets $< F, K, V >$. We also extend the signal restriction operator to triplets:

$$< F, K, V > \setminus S =_{def} < F \setminus S, K, V > .$$

We also define the similar shared variable restriction operator:

$$< F, K, V > \backslash s =_{def} < F, K, V \setminus \{s\} > .$$

For presentation reasons, we shall use both vertical and horizontal presentation for the potential function triplets:

$$< F, K, V > = \left\langle \begin{array}{c} F \\ K \\ V \end{array} \right\rangle$$

5.5.1 Reduction to Non-Dotted Terms

The reduction to non-dotted terms is based on the elementary dot-removal rule:

1. $Can(\bullet \overline{p}, E) = Can(\overline{p}, E)$

The remaining dot-removal rules only take argument terms over derived statements. In these cases, the potential computation is reduced to potential computations over sub-statement terms. The recursive decomposition process ends with an application of the elementary dot-removal rule.

The rules for tests are the simplest ones; the behavior of such statements is that of the branch containing control.

2. $Can(\text{present } S \text{ then } \dot{p} \text{ else } q \text{ end}, E) = Can(\dot{p}, E)$

3. $Can(\text{present } S \text{ then } p \text{ else } \dot{q} \text{ end}, E) = Can(\dot{q}, E)$

4. $Can(\text{suspend } \dot{p} \text{ when } S, E) = Can(\dot{p}, E)$

5. $Can(\text{if } v \text{ then } \dot{p} \text{ else } q \text{ end}, E) = Can(\dot{p}, E)$

6. $Can(\text{if } v \text{ then } p \text{ else } \dot{q} \text{ end}, E) = Can(\dot{q}, E)$

More complicated rules handle sequencing. For instance, when control is in the first branch we have to compose the potential of the two branches:

7. $Can(\dot{p} \,;\, q, E)$

$$= \begin{cases} Can(\dot{p}, E) & \text{if } 0 \notin Can_K(\dot{p}, E) \\ \left\langle \begin{array}{c} Can_S(\dot{p}, E) \cup Can_S(q, E) \\ (Can_K(\dot{p}, E) \setminus \{0\}) \cup Can_K(q, E) \\ Can_V(\dot{p}, E) \cup Can_V(q, E) \end{array} \right\rangle & \text{if } 0 \in Can_K(\dot{p}, E) \end{cases}$$

8. $Can(p \,;\, \dot{q}, E) = Can(\dot{q}, E)$

The potentials of parallel branches are composed using the following rule.

9. $Can(\dot{p}\ ^\perp||^\perp\ \dot{q}, E) = \left\langle \begin{array}{c} Can_S(\dot{p}, E) \cup Can_S(\dot{q}, E) \\ \max(Can_K(\dot{p}, E), Can_K(\dot{q}, E)) \\ Can_V(\dot{p}, E) \cup Can_V(\dot{q}, E) \end{array} \right\rangle$

10. $Can(\dot{p}\ ^\perp||^k\ q, E) = <Can_S(\dot{p}, E), \max(Can_K(\dot{p}, E), \{k\}), Can_V(\dot{p}, E)>$

11. $Can(p\ ^k||^\perp\ \dot{q}, E) = <Can_S(\dot{q}, E), \max(\{k\}, Can_K(\dot{q}, E)), Can_V(\dot{q}, E)>$

12. $Can(p\ ^k||^l\ q, E) = <\emptyset, \{\max(k, l)\}, \emptyset>$

The rule for loop is derived from the rule of the sequence. If the body can terminate and $m = go$, the potential of the loop also includes the potential of the second incarnation of the body.

13. $Can(\texttt{loop}^m\ \dot{p}\ \texttt{end}, E)$

$= \begin{cases} \left\langle \begin{array}{c} Can_S(\dot{p}, E) \cup Can_S(p, E) \\ (Can_K(\dot{p}, E) \backslash \{0\}) \cup Can_K(p, E) \\ Can_V(\dot{p}, E) \cup Can_V(p, E) \end{array} \right\rangle & \text{if } \begin{cases} 0 \in Can_K(\dot{p}, E) \\ \text{and} \\ m = go \end{cases} \\ Can(\dot{p}, E) & \text{otherwise} \end{cases}$

The potential of the started signal declaration statement is that of its body, when the environment is enriched with the current status of the declared signal.

14. $Can(\texttt{signal}\ S^m\ \texttt{in}\ \dot{p}\ \texttt{end}, E) = Can(\dot{p}, E * S^m) \setminus S$

Note how this is simpler than the non-dotted rule on page 64. There, we needed to perform signal feedback at the level of the signal declaration within the potential computation. Here, specific microsteps determine signal absence perform signal feedback.

The other scope declarations also need simple rules:

16. $Can(\texttt{trap}\ T\ \texttt{in}\ \dot{p}\ \texttt{end}, E) = <Can_S(\dot{p}, E), \downarrow Can_K(\dot{p}, E), Can_V(\dot{p}, E)>$

17. $Can(\texttt{var}\ v\ \texttt{in}\ \dot{p}\ \texttt{end}, E) = Can(\dot{p}, E)$

18. $Can(\texttt{shared}\ s\ \texttt{in}\ \dot{p}\ \texttt{end}, E) = Can(\dot{p}, E) \setminus s$

5.5.2 Non-Dotted Terms over Dataless Primitives

Given that the Can_S and Can_K components of these functions have been defined for the constructive behavioral semanticsin Section 4.2. Here, we only define the shared variable component Can_V.

19. $Can_V(\texttt{pause}, E) = Can_V(\widehat{\texttt{pause}}, E) =$

$$Can_V(\texttt{emit}\ S, E) = Can_V(\texttt{exit}\ T(k), E) = \emptyset$$

20. $Can_V(\text{signal } S \text{ in } \overline{p} \text{ end}, E)$

$$= \begin{cases} Can_V(\overline{p}, E * S^\perp) & \text{if } S^1 \in Can_S(\overline{p}, E * S^\perp) \\ Can_V(\overline{p}, E * S^0) & \text{otherwise} \end{cases}$$

21. $Can_V(\text{present } S \text{ then } p \text{ else } q \text{ end}, E)$

$$= \begin{cases} Can_V(p, E) & \text{if } S^1 \in E \\ Can_V(q, E) & \text{if } S^0 \in E \\ Can_V(p, E) \cup Can(q, E) & \text{if } S^\perp \in E \end{cases}$$

22. $Can_V(\text{present } S \text{ then } \widehat{p} \text{ else } q \text{ end}, E) = Can_V(\widehat{p}, E)$

23. $Can_V(\text{present } S \text{ then } p \text{ else } \widehat{q} \text{ end}, E) = Can_V(\widehat{q}, E)$

24. $Can_V(\text{suspend } p \text{ when } S, E) = Can_V(p, E)$

25. $Can_V(\text{suspend } \widehat{p} \text{ when } S, E) = \begin{cases} \emptyset & \text{if } S^1 \in E \\ Can_V(\widehat{p}, E) & \text{otherwise} \end{cases}$

26. $Can_V(\text{trap } T \text{ in } \overline{p} \text{ end}, E) = Can_V(\overline{p}, E)$

27. $Can_V(p \parallel q, E) = Can_V(p, E) \cup Can_V(q, E)$

28. $Can_V(\widehat{p} \parallel \widehat{q}, E) = Can_V(\widehat{p}, E) \cup Can_V(\widehat{q}, E)$

29. $Can_V(\widehat{p} \parallel q, E) = Can_V(\widehat{p}, E)$

30. $Can_V(p \parallel \widehat{q}, E) = Can_V(\widehat{q}, E)$

31. $Can_V(\overline{p} \, ; q, E) = \begin{cases} Can_V(\overline{p}, E) \cup Can_V(q, E) & \text{if } 0 \in Can_K(\overline{p}, E) \\ Can_V(\overline{p}, E) & \text{otherwise} \end{cases}$

32. $Can_V(p \, ; \widehat{q}, E) = Can_V(\widehat{q}, E)$

33. $Can_V(\text{loop } p \text{ end}, E) = Can_V(p, E)$

34. $Can_V(\text{loop } \widehat{p} \text{ end}, E)$

$$= \begin{cases} Can_V(\widehat{p}, E) \cup Can_V(p, E) & \text{if } 0 \in Can_K(\widehat{p}, E) \\ Can_V(\widehat{p}, E) & \text{otherwise} \end{cases}$$

5.5.3 Non-Dotted Terms over Data-Handling Primitives

We complete the definition of the potential function with its expression on the newly-introduced data-handling primitives.

35. $Can(v := f(v^1, \ldots, v^n, s^1, \ldots, s^m), E) = \;<\emptyset, \{0\}, \emptyset>$
36. $Can(s <= f(v^1, \ldots, v^n, s^1, \ldots, s^m), E) = \;<\emptyset, \{0\}, \{s\}>$
37. $Can(\text{var } v \text{ in } \overline{p} \text{ end}, E) = Can(\overline{p}, E)$
38. $Can(\text{if } v \text{ then } p \text{ else } q \text{ end}, E) = Can(p, E) \cup Can(q, E)$
39. $Can(\text{if } v \text{ then } \widehat{p} \text{ else } q \text{ end}, E) = Can(\widehat{p}, E)$
40. $Can(\text{if } v \text{ then } p \text{ else } \widehat{q} \text{ end}, E) = Can(\widehat{q}, E)$
41. $Can(\text{shared } s \text{ in } \overline{p} \text{ end}, E) = Can(\overline{p}, E) \setminus s$

5.6 Behaviors as Sequences of Microsteps

To define the behavior of a statement, we need to chain microstep transitions into reactions. The chaining process must respect both the constructive synchronization of the signals and the classical causality of data accesses. We define a new relation describing the composition of microsteps:

$$\dot{\overline{p}}, data \xrightarrow[E]{E',k}{}^* \ddot{\overline{p}}, data'$$

which holds, by definition, when there exists a sequence

$$\dot{\overline{p}}, data \xrightarrow[E]{e_1,\perp} (\dot{\overline{p}})_1, data_1 \xrightarrow[E]{e_2,\perp} \ldots \xrightarrow[E]{e_{n-1},\perp} (\dot{\overline{p}})_{n-1}, data_{n-1} \xrightarrow[E]{e_n,k} \ddot{\overline{p}}, data'$$

with $E' = \{e_i \mid e_i \neq \perp\}$. If the sequence of rewriting steps that defines $\dot{\overline{p}}, data \xrightarrow[E]{E',k}{}^* \ddot{\overline{p}}, data'$ is maximal in the given signal context E, we also write

$$\dot{\overline{p}}, data \xrightarrow[E]{E',k} \!\!\!\!\!\twoheadrightarrow \ddot{\overline{p}}, data'.$$

We are now able to define the notion of constructiveness in the new operational framework (effectively extending the notion to programs with data).*

Definition 2 *A statement p in the (macrostep) state $\overline{p}$ is constructive in the context E and for the store data if there exist $\overline{p}'$, $data'$, E', and $k \neq \perp$ such that*

$$\bullet\overline{p}, data \xrightarrow[E]{E',k} \!\!\!\!\!\twoheadrightarrow \overline{p}', data'$$

---

*Recall that in this book *correct* is synonymous with *constructive*.

The program is *non-constructive* if the resulting term contains bullets. This corresponds to the situation where control blocked during the execution because the status of a signal or shared variable cannot be established. We describe the relation between the CBS and COS definitions of constructiveness in Section 5.7.

Determinism of the COS

While the result is not fully proven, we assume the COS of Esterel is deterministic, as stated by the following conjecture.

Conjecture 1 (Rewriting confluence) *For any term $\dot{p}$ and for any environment E we have*

$$\left. \begin{array}{l} \dot{p}, data \xrightarrow[E]{E'_1, k_1} (\dddot{p})_1, data_1 \\ \dot{p}, data \xrightarrow[E]{E'_2, k_2} (\dddot{p})_2, data_2 \end{array} \right\} \Rightarrow \left\{ \begin{array}{l} k_1 = k_2 \\ E'_1 = E'_2 \\ (\dddot{p})_1 = (\dddot{p})_2 \\ data_1 = data_2 \end{array} \right.$$

As its name suggests, the determinism result is actually a confluence result that concerns the result of maximal transition sequences. The sequences themselves do not need to be identical, and non-identical derivations can be easily built, because parallel branch concurrency is modeled by non-deterministic interleaving, and deciding the absence of a signal (using the *Can* function) can be usually done at several different points during a derivation.

The previous conjecture is intuitive, given the construction of the COS. However, it is difficult to prove because the complexity of proving the confluence of term rewriting is augmented by the necessity of proving the confluence of the side effects on the data store; and the proof involves a large number of (usually simple) cases, corresponding to the different statements and to the corresponding transition rules.

Program behavior

In Section 4.7, we explained how the behavior of a program is defined based on the behavior of its body. We also explained why certain restrictions must be satisfied by the program interface to relate the behaviors of the program and of its body and we gave the semantics-preserving program transformations that ensure these restrictions are met.

This presentation was done with the constructive behavioral semantics, but the approach is easily extended to cope with data in the COS. In fact, the only difference concerns the valued signals and sensors of the interface. These signals, or the associated primitive (kernel) objects—the interface shared variables—must be treated the same as pure signals. Connection code must be added in cases where an interface signal or shared variable can be both read and emitted in an instant.

When the needed connection code is added, we define a new program behavior relation as follows. Let P be a Kernel Esterel program of body p that cannot both test and emit an interface signal or shared variable during a reaction. Assume I is the set of input signals of P and O is the set of output signals. Let E_I be an event over I (an input event) and E_O an event over O. Then

$$P', data' \xrightarrow[E_I]{E_O} P'', data'' \Leftrightarrow$$
$$\exists k \in \{0,1\}, \exists E \text{ such that } \bullet\overline{p}', data' \xrightarrow[E_I]{E,k} \overline{p}'', data'' \text{ and } E/O = E_O$$

where $\overline{p}'$ (resp. $\overline{p}''$) is the state of p in P' (resp. P'') and E ranges over events over $I \cup O$. The initial status of the data store must be consistent with the synchronization protocol on input and output shared variables. The status of input variables must be *ready*, and the status of output variables must be *old*.

5.7 COS versus CBS

In this section we discuss the differences between between the two flavors of direct constructive semantics of Esterel. Naturally, the comparison can only be made on Pure Esterelsince the constructive behavioral semantics cannot handle data. This allows us to simplify the COS notation by discarding all references to data in the semantic terms, transition rules, and potential functions.

On Pure Esterel, microstep transitions are written $\dot{\overline{p}} \xrightarrow[E]{e,k} \ddot{\overline{p}}$, and the rules describing composition of microsteps have the forms $\dot{\overline{p}} \xrightarrow[E]{E',k}{}^* \ddot{\overline{p}}$ and $\dot{\overline{p}} \xrightarrow[E]{E',k} \ddot{\overline{p}}$. The potential function is

$$Can(\ddot{\overline{p}}, E) = \langle\, Can_S(\ddot{\overline{p}}, E)\,,\, Can_K(\ddot{\overline{p}}, E)\, \rangle.$$

With these simplifications, we are ready to state that the Constructive Behavioral and the COS are equivalent on Pure Esterel. The following conjectures state that we introduced a language extension, but have not created a completely new language. A Pure Esterel program has the same meaning in the COS as in the constructive behavioral semantics.

As for the determinism results of the previous section, we do not have a proof of our statements. We therefore present them as conjectures.

Conjecture 2 (Equivalence to reference semantics) *For any term $\overline{p}$, completion code $k \neq \bot$, and events E and E', we have*

$$\overline{p} \xrightarrow[E]{E',k} \overline{p}' \Leftrightarrow \bullet\overline{p} \xrightarrow[E]{E',k} \overline{p}'.$$

The proof of the result makes use of a series of intermediate results of which we mention only one, for it gives insight in the strong relation existing between execution and potential computations.

Conjecture 3 (Potential function characterization) *If* $\bullet \overline{p} \xrightarrow[E]{E',k} \dddot{\overline{p}}$, *then*

$$Can(\overline{p}, E) = Can(\bullet\overline{p}, E)$$

$$Must(\overline{p}, E) = \langle\, E'\,,\, \{k\}\backslash\{\bot\}\,\rangle$$

$$Can^+(\overline{p}, E) = \begin{cases} Must(\overline{p}, E) \cup Can(\dddot{\overline{p}}, E) & \text{if } k = \bot \\ Must(\overline{p}, E) & \text{if } k \neq \bot \end{cases}$$

The result emphasizes the operational aspect of our semantics where potential analysis is restricted to code pruning.

6
Constructive Circuit Translation

In Chapter 3 we explained the semantic principles that allow us to represent a Pure Esterel program as a constructive logic formula. We also explained that ternary circuit simulation provides an operational way of determining the solutions of such formulas. These solutions determine the reactions, i.e., the behavior of the program.

In this chapter, we explain how the basic constructive circuit model is extended to represent the data-handling primitives of Esterel, and we then show how Esterel programs can be translated into such Boolean circuits. Thus, we obtain a circuit version of the source behavior amenable to hardware synthesis and software generation; and an entry point to a large class of circuit-level verification, simulation, and optimization techniques.

In the next chapter (Section 7.3), we present the INRIA compiler, whose software code generation schemes are based on the simulation of the resulting circuits. As the following chapters will show, circuits with data form the semantic domain where the formal semantics of the GRC compilation format is given. Stating or proving the correctness of GRC-level optimization and code generation techniques relies on a fine understanding of the circuit model.

This chapter is divided into two parts. In the first, we present the notion of *constructive circuit with data*, which was introduced by the Esterel Group [32] to partially extend the use of circuit-level techniques to the full Esterel language. In circuits with data, the evaluation of wires may trigger *data actions*, which are routines that read and assign data variables. Some actions are tests whose Boolean return value can be taken into account in the computation of wire values.

Although circuits with data can represent non-deterministic behavior, by construction, circuits obtained from constructive Esterel programs are deterministic. To achieve determinism, these circuits employ data dependencies to ensure an ordering of actions consistent with the desired side effects.

In the second part of this chapter, we present the circuit translation scheme in detail. The main principles behind the circuit translation of Esterel have not changed since they were first defined by Berry [6, 7].

Control flow and code pruning are represented in the Boolean circuit by the propagation of 1 and 0 values respectively.

The control state of the program between execution instants (the status of its **pause** statements) is stored in Boolean registers, i.e., memory elements whose output at a cycle is the input at previous cycle.

The circuit translation uses wires for signals and control paths and registers for **pause** statements. This approach allows the definition of simple, structural translation scheme where the circuit associated with a statement is obtained by composing the circuits of its sub-statements into predefined patterns.

Many semantically-equivalent circuit translations can be defined that are consistent with these guidelines. Among them, the translation presented in this book has been designed for simple presentation. More precisely, we use the digital circuit representation to connect the direct COS to intermediate compilation representations. The components of a generated circuit correspond to the control flow structures of the COS. Moreover, the GRC intermediate representation of an Esterel program (defined in Chapter 8) can be seen as an abstraction of the corresponding circuit. Its formal semantics are given in the circuit with data domain. In Appendix B, we explain how more complex translations support experimental language primitives.

6.1 Digital Circuits with Data

Esterel programs are translated into digital circuits at the *logic gate level*. At this level, the circuits are composed of two types of objects:

- *combinational gates* (AND, OR, NOT, etc.), which are the elementary components in the computation of a reaction; and

- *Boolean registers*, which store the status of the program between reactions.

Gates and registers are connected by wires that carry values from a source gate to an arbitrary number of destination gates. A circuit for ABRO (an example first presented on page 7) is given in Figure 6.1.

The dotted boxes show the gates for each statement. The leftmost box corresponds to the preemption test for R; the rightmost one to the parallel synchronization; the two remaining boxes correspond to the "await A" and "await B" statements.

6.1.1 Circuit Semantics. Constructive Causality

To execute Esterel programs, digital circuits are endowed with *constructive semantics* [7], based on a constructive value propagation model. There are two equivalent ways to present constructive evaluation of gates (see Berry [7]).

```
module ABRO:
input A, B, R ;
output O ;
loop
  [
    await A
  ||
    await B
  ];
  emit O
each R
end module
```

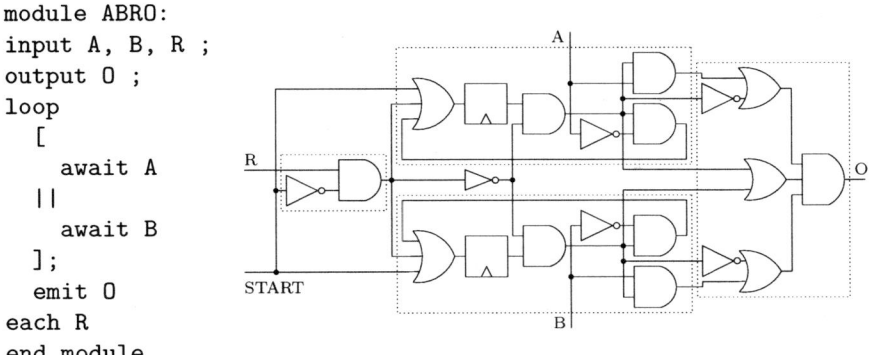

Figure 6.1: A possible circuit translation for ABRO

The first way is to use logical rewrite rules such as $1 \vee x \to 1$ and $1 \vee x \to x$. Since only constructive rules are considered, the excluded midde law $x \vee \neg x = 1$ is not applicable. The second way is to compute wire values in the 3-valued model $\mathcal{B}_\bot = \{\bot, 0, 1\}$, with the same rules as above where x is a value in $\mathcal{B}_\bot$. These rules first appeared on page 46.

x	y	$x \wedge y$	$x \vee y$	$\neg x$
$\bot$	$\bot$	$\bot$	$\bot$	$\bot$
$\bot$	0	0	$\bot$	$\bot$
$\bot$	1	$\bot$	1	$\bot$
0	$\bot$	0	$\bot$	1
0	0	0	0	1
0	1	0	1	1
1	$\bot$	$\bot$	1	0
1	0	0	1	0
1	1	1	1	0

Note that a gate can compute its result based on incomplete input information (for instance $\bot \vee 1 = 1$). These approaches are equivalent, as shown in Berry [7]. Here, we take the second approach.

We assume the values on the input wires are defined at the beginning of the reaction and hold throughout. Constructive value propagation works as follows. We first mark all non-input wires with $\bot$. This corresponds to when no gate has performed any computation. Then, we propagate the values of gates that have enough input information to compute a defined output value. The values are always mechanically constructed from inputs, never guessed. In this process, gate outputs can only change from $\bot$ to 0 or 1, never from 0 to 1 or 1 to 0, meaning the computation is monotonic. Since the number of wires is finite, the computation reaches a fixpoint after a finite number of

106 CONSTRUCTIVE CIRCUIT TRANSLATION

Figure 6.2: The two possible evaluation sequences for a small circuit

Figure 6.3: Causality in circuit evaluation

gate evaluations. This fixpoint is the constructive result of the circuit for the given inputs.

An essential property is that the fixpoint is independent of the order in which gates are evaluated. Figure 6.2 presents two evaluation sequences for a small circuit with the same inputs, illustrating how the results is independent of evaluation order. For simplicity, we assume in the sequel that the evaluation of the circuit is performed gate by gate, and we shall call the evaluation of a gate an evaluation step.

Another important property is that wire dependencies can order the evaluation of different wires. For instance, only one evaluation sequence exists for the input valuation of Figure 6.3. Indeed, the AND gate must await the computation of its second argument to start. This is the constructive causality we will exploit in the next section where we define circuits with data.

Sequential behavior

In Esterel, the **pause** statement is responsible for dividing the program behavior into successive reactions. The Boolean register plays this role in digital circuits.

A Boolean register is a circuit component with one input wire (IN), one output wire (OUT), and an initial value (V).

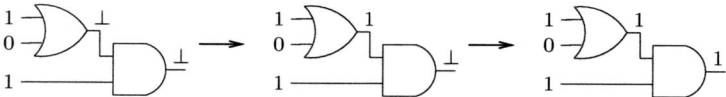

If IN_i and OUT_i are the values of the input and output wires during the

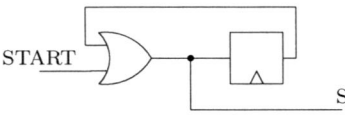

Figure 6.4: A simple sequential circuit

ith reaction (clock cycle), then the behavior of the register is defined by

$$\text{OUT}_i = \begin{cases} V, & \text{if } i = 0 \\ \text{IN}_{i-1} & \text{otherwise.} \end{cases}$$

Most registers used in the translation of Esterel into digital circuits have an initial value of 0, so we will not bother to write the initial value on most registers.

The translation of an Esterel statement into a digital circuit is performed by using one Boolean register for each **pause** statement. For instance, consider the kernel expansion of "sustain S".

```
loop
   emit S ;
   pause
end
```

In the generated circuit, we use a 1 to indicate both execution and signal presence. Under this encoding, Figure 6.4 is one way to implement this program. We set the START wire to 1 in the first reaction and set it to 0 in all others. The program resumes from the **pause** statement when the corresponding register output is 1. When the program is started or when the **pause** statement is resumed, the signal S is emitted by setting the corresponding wire to 1 and the **pause** statement is activated by setting its register input is set to 1.

More complex examples will be presented later. In general, the translation of **pause** wraps its register a sub-circuit that freezes or preempts the control flow as required by enclosing preemption structures.

6.1.2 Extension to Circuits with Data

Data actions

A major advantage of the constructive evaluation approach is that it can be extended to support data operations, necessary for handling the full Esterel language. To see how, consider adding a signal test and some data handling code to the previous example.

```
var v := 0 : integer in
   loop
```

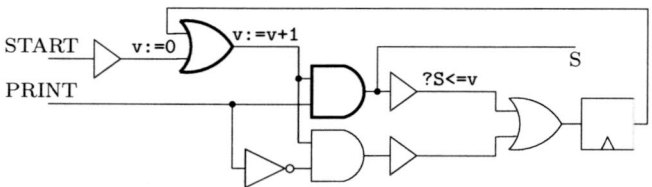

Figure 6.5: Circuit with data, first example

```
      v := v + 1 ;
      present PRINT then emit S(v) end ;
      pause
    end
  end
```

This can translated into the circuit with data of Figure 6.5. The protocol on START is the same as before. The signal S is emitted only in reactions where PRINT is present. The data actions "v:=0," "v:=v+1," and "?S<=v" are performed during reaction evaluation whenever their associated wire becomes 1 and before evaluation propagates this value further. By this mechanism, constructive evaluation orders actions in instants where both are executed. For instance, in reactions where control resumes from the register and PRINT is present, the identity gates controlling the data actions are evaluated as shown in Figure 6.3, and the assignment is performed before the test.

Data dependencies

For valued signals, the constructive evaluation of the control flow wires are not enough to ensure a correct scheduling of the data actions. Consider the following Esterel fragment.

```
      ?S <= 10 || v := ?S
```

When this is started, the shared variable assignment must be executed before the sequential variable v is assigned to guarantee that the current value of S is assigned to v to match the Esterel semantics. Since the data actions belong to different branches of the parallel, the constructive evaluation of the control wires in Figure 6.6(a) is not sufficient to ensure the correct ordering of the actions. We need to introduce additional *data dependency* arcs, drawn as dashed arrows. Such arcs mark scheduling constraints due to data and not implied by the constructive evaluation of the regular circuit wires.

Although for convenience, we have explicit data dependencies (i.e., the dotted lines) in our formalism, it is not strictly necessary. The same effect can be achieved with additional standard gates. For example, Figure 6.6(b) shows how to replace the dotted data-dependency link by three additional

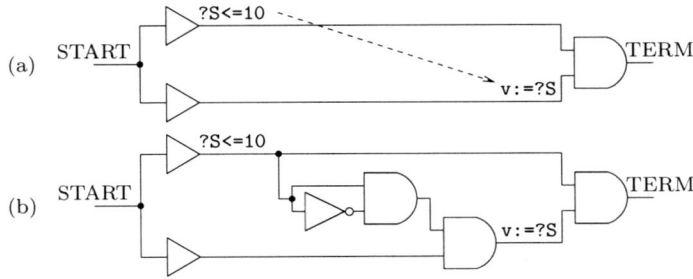

Figure 6.6: Circuit with data, second example

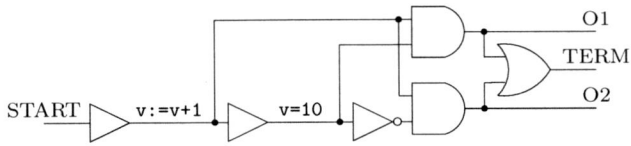

Figure 6.7: Circuit with data, third example

gates that propagate an auxiliary 1 and together imply the same dependency. The extra gates allow the action "v:=?S" to be started only when its trigger is set to 1, i.e., after "?S<=10" executes or when we have determined that it will not.

Test actions

Circuits with data not only drive the execution of data actions, but may also make decisions based on data tests. When gate driving a test action becomes 1, the action is executed, and the output wire is set to the Boolean result value of the test. In other words, a wire with a test action may actually have a different output value than its input. Figure 6.7 shows the circuit for the following example, which includes the data test v = 10.

```
v := v + 1;
if v = 10 then emit O1 else emit O2 end
```

Data abstraction issues

In the circuit-with-data representation, we separate data from control and do not generate gates for data computations. This common abstraction enables much more aggressive analysis. For instance, Sentovich et al. [63] show how to perform sequential optimizations on the control circuit, which would be impossible on the full circuit with data since the state space would immediately explode. In practice, a reasonable strategy for synthesizing a circuit in which gates also handle data is to optimize the control circuit separately, then sew it together with circuitry for the datapath. The Esterel V7 compiler does this.

6.1.3 Formal Definitions

A digital circuit with data $\mathcal{C}$ is a set $\mathcal{W}$ of wires, a subset $\mathcal{I} \subseteq \mathcal{W}$ of input wires, a set $\mathcal{V}$ of variables, gate definitions, and a set $Causal$ of explicit causal dependencies.

The set of wires contains the circuit inputs, whose value are provided by the environment, and the output wire of each gate. The output wire $w \in \mathcal{W} \setminus \mathcal{I}$ of each gate has a constructive logic expression over the values of the other wires, denoted $f_w : \mathcal{B}_\perp^W \to \mathcal{B}_\perp$. The function f_w is monotonic with respect to $\mathcal{B}_\perp$.

Each variable v has a type $Type(v)$. To simplify the evaluation algorithm, we assume that every wire w triggers one action act_w when set to 1. Based on the current status of the variables, act_w may assign new values to some variables, as defined by the associated function $val_w : \prod_{v \in \mathcal{V}} Type(v) \to \prod_{v \in \mathcal{V}} Type(v)$. The call to act_w also produces a Boolean value with the test function $test_w : \prod_{v \in \mathcal{V}} Type(v) \to \mathcal{B}$. Intuitively, the action of a wire w is a test action when $test_w$ is not constant 1. The set $Causal$ of causal action ordering relations is composed of pairs of wires. If $(w_1, w_2) \in Causal$, then the evaluation of w_1 must be performed before the evaluation of w_2.

Because we are mostly interested in the evaluation of a single reaction, our definitions only concern the combinational part of the circuit. We represent registers as a pair of wires: one wire for the output of the register and one for the input.

An evaluation of the circuit $\mathcal{C} = \langle \mathcal{W}, \mathcal{I}, \mathcal{V}, Causal \rangle$ is a sequence of circuit valuations $\dot{\mathcal{C}}_i = \langle \dot{\mathcal{W}}_i, data_i \rangle$, $1 \le i \le n$, where $\dot{\mathcal{W}}_i : \mathcal{W} \to \mathcal{B}_\perp$ and $data_i$ are stores representing the status of all variables ($data_i(v) \in Type(v)$). At each step of the evaluation process one gate is evaluated and one wire changes its status from $\perp$ to either 0 or 1. If the ith evaluation step corresponds to the evaluation of wire w, we write $\dot{\mathcal{C}}_i \xrightarrow{w} \dot{\mathcal{C}}_{i+1}$ for the state transition.

The evaluation steps are subject to causal restrictions. If $\dot{\mathcal{W}}_i(w) = \perp$ and $f_w(\dot{\mathcal{W}}_i) \ne \perp$, we can evaluate w during the ith evaluation step if either

- $f_w(\dot{\mathcal{W}}_i) = 0$, in which case $data_{i+1} = data_i$ and

$$\dot{\mathcal{W}}_{i+1}(w') = \begin{cases} \dot{\mathcal{W}}_i(w') & \text{if } w' \ne w \\ 0 & \text{if } w' = w; \end{cases} \quad \text{or}$$

- $f_w(\dot{\mathcal{W}}_i) = 1$ and all the explicit causal dependencies are satisfied (for all $(w', w) \in Causal$, we have $\dot{\mathcal{W}}_i(w') \ne \perp$). In this case,

$$data_{i+1} = val_w(data_i)$$

$$\dot{\mathcal{W}}_{i+1}(w') = \begin{cases} \dot{\mathcal{W}}_i(w') & \text{if } w' \neq w \\ test_w(data_i) & \text{if } w' = w. \end{cases}$$

An evaluation of the circuit $\mathcal{C}$ is any sequence of evaluation steps.

$$\dot{\mathcal{C}}_1 \xrightarrow{w_1} \dot{\mathcal{C}}_2 \xrightarrow{w_2} \cdots \xrightarrow{w_n} \dot{\mathcal{C}}_{n+1}$$

By construction, any evaluation sequence is finite. For a general circuit with data, the evaluation process is not necessarily confluent, i.e., does not lead to a uniquely defined state, largely because side-effects may be non-commutative. However, our translation process for Esterel produces a confluent circuit by adding data dependencies to avoid this problem.

Constructiveness of dataless circuits

Note that classical dataless circuits can be represented in our model. A circuit in our model is dataless when its variable and causal dependency sets are empty and all test functions are constant 1. In this case, we can simplify the notation. First of all, the variable and causal dependency sets can be omitted from the circuit tuple $\mathcal{C} = \langle \mathcal{W}, \mathcal{I} \rangle$. Furthermore, a circuit valuation coincides with the valuation of its wires $\dot{\mathcal{C}} = \langle \dot{\mathcal{W}} \rangle$.

When considering dataless circuits, we fall onto the classical definitions of constructiveness as given in Section 3.1.1. The monotonicity of the logic gate functions f_w ensures that the evaluation of the circuit is monotonic and confluent for any given input. The circuit is constructive when for any input valuation the ternary simulation associates a 0 or 1 value with every circuit wire.

Definition 3 (dataless constructiveness) *Let $\mathcal{C} = \langle \mathcal{W}, \mathcal{I} \rangle$ be a dataless circuit and let $\dot{\mathcal{W}}$ be an initial valuation that assigns 0 or 1 values to all inputs and $\bot$ with all other wires. $\mathcal{C}$ is* constructive for the initial valuation $\dot{\mathcal{W}}$ *when any maximal evaluation of $\mathcal{C}$ starting in $\dot{\mathcal{W}}$ assigns a value of 0 or 1 to every wire. $\mathcal{C}$ is* constructive *when it is constructive for every initial valuation $\dot{\mathcal{W}}$.*

When the dataless circuit is associated with a Pure Esterel program, its constructiveness means it can compute the reaction of the program in any state and for any input.

Figure 6.8(a) shows a dataless non-constructive circuit. When the input is set to 1, the output cannot be constructively assigned a value of 0 or 1. The circuit corresponds to the following Pure Esterel statement.

```
present S then emit S end
```

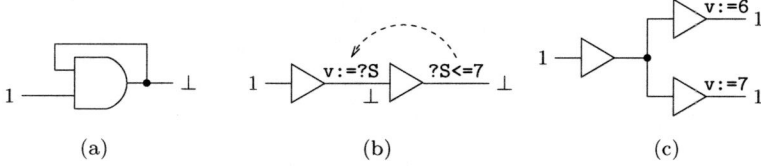

Figure 6.8: Incorrect circuits

Constructiveness of circuits with data

Defining constructiveness is far more complex for circuits with data. We expect termination and determinism from circuits obtained from Esterel programs.

Definition 4 (Termination, Determinism, and Constructiveness)
Let $\mathcal{C} = \langle \mathcal{W}, \mathcal{I}, \mathcal{V}, Causal \rangle$ *be a circuit with data, let d be a valuation of the variables, and let $\dot{\mathcal{W}}$ be the initial values of the wires.*

1. $\mathcal{C}$ *terminates for the initial values* $\dot{\mathcal{C}} = \langle \dot{\mathcal{W}}, d \rangle$ *when any maximal circuit evaluation starting in* $\dot{\mathcal{C}}$ *assigns a value of 0 or 1 to every wire.*

2. $\mathcal{C}$ *is deterministic, or confluent, for the initial values* $\dot{\mathcal{C}}$ *when all maximal circuit evaluations starting in* $\dot{\mathcal{C}}$ *finish with the same wire and store values.*

3. $\mathcal{C}$ *is constructive for the initial values* $\dot{\mathcal{C}}$ *if it is both deterministic and terminating for those values.*

All dataless circuits are deterministic, and all constructive dataless circuits are constructive for any input valuation. When data is involved, things become more complicated. Figure 6.8(b) presents a circuit that is deterministic for any input valuation, but does not terminate when its input is set to 1. The circuit corresponds to the Esterel program

```
v := ?S ; emit S(10)
```

Figure 6.8(c) presents a circuit that terminates for any input but is not deterministic when the input wire is set to 1.

Later, we will require circuits generated from constructive Esterel programs to be constructive for consistent states of the program.

6.2 Translation Principles

An Esterel program is translated into a Boolean circuit structurally. Each statement of the program is associated with a small circuit. The circuit for a composite statement comes from composing the circuits of its sub-statements. The translation of a program is performed in a bottom-up fashion, starting at simple statements emit, pause, and exit.

There is one pattern for each kernel Esterel statement. Derived statements can first be dismantled into kernel statements, but in practice using special patterns for derived statements produces better circuits. All compilers take this route, but we will not provide the details here.

Our circuit translation reuses the three main notions of the Esterel operational semantics:

1. *Selection* is the notion describing the state encoding of Esterel. A statement is *selected* for execution in an instant if control paused on an enclosed pause in the previous instant. Thus, pause statements are the basic Boolean state holders. In the circuit, selection will be derived by OR-ing selection wires out of the pause registers.

2. The *start behavior*, also called *surface behavior*, is the computation of a statement when the microstep control flow reaches it while it is not selected. The start behavior does not need to read the selection statuses of its sub-statements, for they are known (not selected). Due to the complex phenomenon of *reincarnation*, a statement can be started several times during a single execution instant.

3. The *resumption behavior*, also called *depth behavior*, is the computation of a statement resumed when currently selected. The computation of the depth behavior depends on the computation of the selection statuses of its sub-statements and it may itself lead to start or resumption of these sub-statements. Note that statements such as emit that contain no pause have no state, are never resumed, and therefore exhibit no depth behavior.

A statement can only be resumed once during an execution instant. However, because of loops, the statement p in "loop p end" is resumed and started in the same instant when it loops.* In complex reincarnation cases, a statement can be restarted several times in the same instant, as explained in Section 2.3, page 30.

Our translation scheme will associate with each statement three structurally-defined circuits, corresponding to the three notions listed above. For a statement p, we define

1. the *selection circuit* $CSelect(p)$, which preserves the state of the statement between successive clock cycles and computes the selection statuses of all the (sub-)statements of p;

2. the *surface circuit* $CSurf(p)$, which computes the start behavior of p; and

---
*Note that the loop body is no longer selected when it is restarted.

3. the *depth circuit* $CDepth(p)$, which computes the resume behavior of p and is activated only in instants where the statement is selected.

The surface and depth circuits are combinational. All registers appear in the selection circuit, whose wires depend solely on registers and not on inputs. Therefore, in constructive evaluation, selection gates can be computed before surface and depth gates.

In the remainder of this section, we define the translation interfaces of each of these circuits and we explain how to generate the global program circuit by connecting the selection, surface, and depth circuits for the program body.

6.2.1 The Selection Circuit

Section 6.1 explained how we use Boolean registers to implement **pause** statements. Our translation generates one Boolean register per **pause** statement, plus a boot register for the whole program, corresponding to an implicit leading **pause** selected at first instant.

In addition to the registers, the selection circuit associated with a statement uses a tree of OR gates to compute the selection status of all its sub-statements. Consider the following Esterel program.

```
module Small:
input I,K;
output J;
trap T in
  await I ; pause ; sustain J
||
  await K ; exit T
end trap
end module
```

The selection circuit of Small is pictured in Figure 6.9. The special boot register is associated with the full program. The other registers are generated by **pause** statements: one for the explicit **pause**, and three for the implicit **pause**es: two for **await** and one for **sustain**.

All registers except the boot register have initial value 0, which makes the initial state of the Esterel program the case where the initial **pause** is selected. The boot register outputs 1 in the first reaction of the program and outputs 0 in all later reactions because its input wire is set to 0.

Given a statement p, the inputs of the selection circuit $CSelect(p)$ are the register input wires $RIN(rp)$, where rp ranges over all the registers generated by the **pause** statements of p. When p is a program and rp is its boot **pause**, $RIN(rp)$ is always set to 0. The outputs of $CSelect(p)$ are the $SEL(p)$ wires, where p ranges over all the sub-statements of p, including p itself.

For the Small example, the derived Esterel statements **await** and **sustain** each generate one register. The standard kernel expansion of more complex

TRANSLATION PRINCIPLES   115

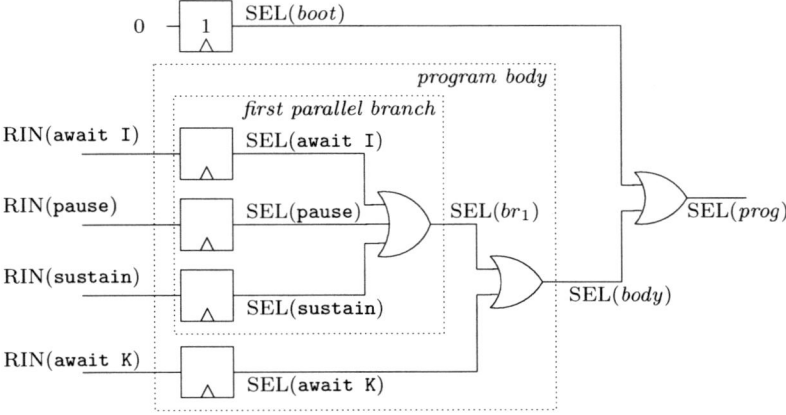

Figure 6.9: The selection circuit of a simple example

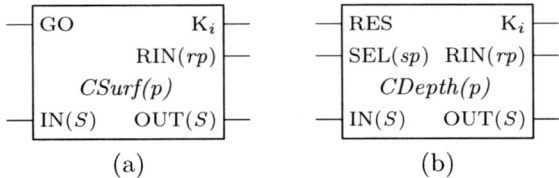

Figure 6.10: The interface of the surface (a) and depth (b) circuits associated to a statement p in our translation.

derived statements such as "**every immediate**" would generate two registers, which is unnecessary. For each derived statement, there is an optimized translation with only one register.

6.2.2 The Surface and Depth Circuits

The surface and depth circuits for a statement implement the function of each reaction using combinational gates. In the next section, we show how they are connected to the state-holding selection circuit to generate the full sequential behavior of an Esterel program.

Consider a statement p. As pictured in Figure 6.10(a), the inputs of *CSurf(p)* are the wires GO and IN(S), where S ranges over the signals that visible from p, excluding its local signals. The GO wire is set to 1 in cycles where p is started to trigger the start behavior. The IN(S) wires collect the status of each signal from the environment.

The outputs of *CSurf(p)* are the wires K_i, RIN(rp), and OUT(S), where i ranges over the set $K_s(p)$ of potential completion codes of p, see definition

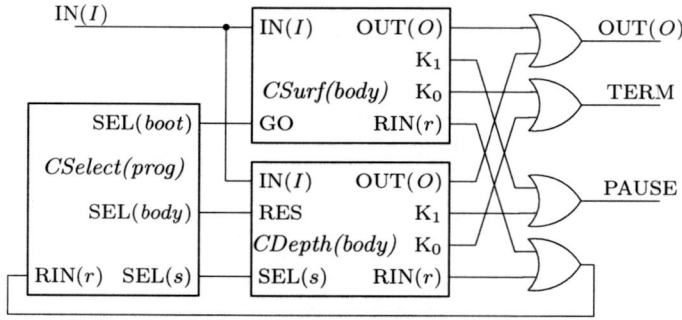

Figure 6.11: The global translation context

in Section 4.6; rp ranges over the **pause** statements of p; and S ranges over signals that can be emitted but are not declared by p.

The K_i wire is set to 1 to represent p's completion with code i. In instants where GO is 1, exactly one of the K_i wires is set to 1. The wire RIN(rp) is the input of the rp selection register; it is set to 1 when control pauses on the corresponding **pause** statement. The wire OUT(S) is set to 1 when the signal is emitted by p. It is set to 0 when p does not emit S.

The depth circuit associated with p is denoted $CDepth(p)$. Its interface (Figure 6.10(b)) is richer than that of $CSurf(p)$. The wires SEL(sp) are inputs that decode the current state of its sub-statements (sp ranges over p's immediate sub-statements). The resumption wire RES is 1 in cycles in which p is resumed.

6.2.3 The Global Context

The circuit representing the full sequential behavior of an Esterel program combines the surface and depth circuits of the program body with the program's selection circuit following the pattern in Figure 6.11. This distributes the inputs to the surface and depth circuits and collects the outputs.

The output SEL(*boot*) of $CSelect(prog)$ is connected to the input GO of $CSurf(body)$; this triggers the first instant, where the implicit **pause** responsible for booting the circuit is selected. In all the other instants, $CSurf(body)$ has value 0. While the program is still active (while SEL(*body*) = 1) the resumption is triggered since SEL(*body*) is connected to RES. The resumption also reads the selection statuses of the different statements of the program, which are computed by $CSelect(prog)$. The status of the registers RIN(r) is computed by $CSurf(body)$ in the start instant and by $CDepth(body)$ in the other instants.

An output signal O is emitted by the program when either $CSurf(body)$ or $CDepth(body)$ sets the corresponding wire OUT(O) to 1. The program

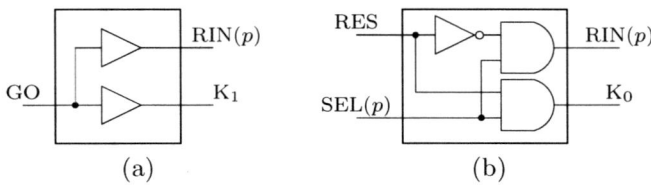

Figure 6.12: Surface (a) and depth (b) circuits for pause

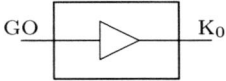

Figure 6.13: Surface circuit for nothing

terminates when the completion code wire K_0 is set to 1 by either the surface or depth circuits.

The global context circuit pictured here assumes that the input and output signal sets are disjoint, meaning there are no inputoutput signals. It also assumes that wrapper code has been added for any input signals that are also emitted by the program (see Sections 4.7 and 5.6).

6.3 Translation Rules

6.3.1 Dataless Primitives

Pause

When the statement $p =$ pause is started by setting the wire GO to 1, the surface circuit (Figure 6.12(a)) sets the register output RIN(p) and the K_1 completion code wires to 1.

The depth circuit (Figure 6.12(b)) produces the completion code K_0 ("terminate") when resumed while being selected (both RES and SEL(p) set to 1). When the statement is selected, but not resumed (i.e., suspended), the register output is set to 1 to preserve the status of the register.

Nothing

The statement nothing is instantaneous and therefore only has a surface circuit: Figure 6.13. The circuit simply copies GO to completion code wire K_1.

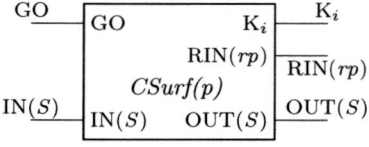

Figure 6.14: Surface circuit for "`loop p end`". The index i ranges over the set $K_s(p)$, which must not contain code 0 (*cf.* Section 4.6).

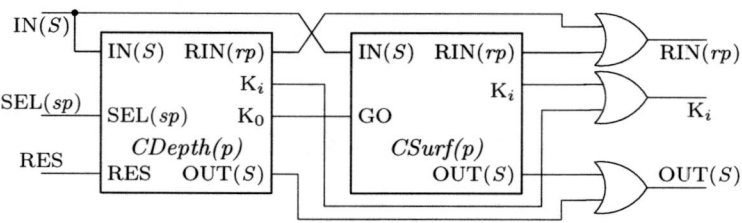

Figure 6.15: Depth circuit for "`loop p end`". The index i ranges over the set $K_d(\texttt{loop } p \texttt{ end})$ (which does not contain 0).

Loop

Starting a loop amounts to starting its body, so the surface circuit of "`loop p end`" is simply the surface circuit of p. Its K_0 is 0 since the body of a loop statement cannot terminate instantaneously.

The depth circuit, Figure 6.15, is more complex because it embodies the solution to the reincarnation problem described in Section 2.3. When resumed, the body of a loop can terminate. In that case it must be instantly restarted. We generate a copy of *CSurf(p)* for the depth of the loop circuit, triggered by the K_0 output of *CDepth(p)*. This copy is specific to the loop depth—distinct from that for the surface. The induced replication of the surface circuit of p is called circuit reincarnation. It ensures that each evaluation of p's surface inside a given execution instant is performed by a fresh instance of either *CDepth(p)* or *CSurf(p)*.

Our translation scheme may replicate code even when doing so is unnecessary. Static analysis techniques like those used in the circuit-based INRIA compiler can reduce the degree of replication, but we shall not investigate this problem here. Instead, we shall exploit the simplicity of the scheme by allowing the definition of a simple link between the circuit translation and the compilation schemes defined in the third part of the book, and facilitating the task of optimization algorithms by producing simpler control and data dependencies.

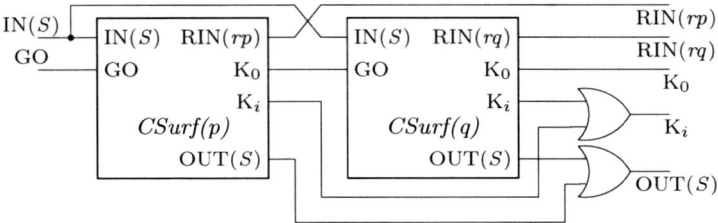

Figure 6.16: Surface circuit for the two-way sequence "$p \; ; \; q$". The index i ranges over $K_s(p \; ; \; q) \setminus \{0\}$.

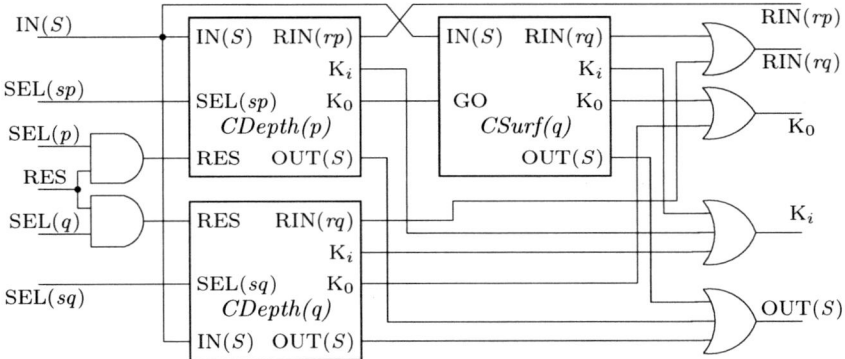

Figure 6.17: Depth circuit for the binary sequence "$p \; ; \; q$" (i ranges over $K_d(p \; ; \; q) \setminus \{0\}$).

Sequence

Figure 6.16 shows the surface circuit for the binary sequence "$p \; ; \; q$." Once started, the sequence starts its first branch by triggering its surface circuit *CSurf(p)*. If p terminates ($K_0=1$), the second branch is started immediately.

Figure 6.17 shows the depth for "$p \; ; \; q$." When resumed, the sequence resumes its active sub-statement, determined by the selection status inputs SEL(p) and SEL(q). When p terminates, we trigger an incarnation of q's surface circuit.

Parallel

Figure 6.18 shows the surface circuit for the parallel statement $p = p_1 \; || \; \ldots \; || \; p_n$. To simplify the figure, we only drew a single sub-circuit (for branch p_j) as well as all the common gates. Common gates are represented using the extensible gate convention used by Berry [7]. We mark extensible gates with

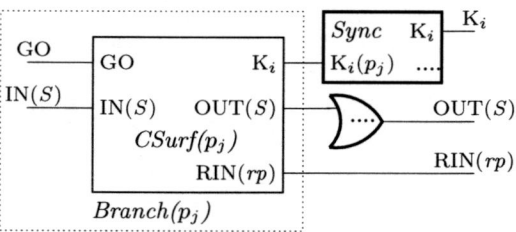

Figure 6.18: Surface circuit for "$p_1 \;||\; \ldots \;||\; p_n$" ($i$ ranges over $\bigcup_{j=0}^{n} K_s(p_j)$)

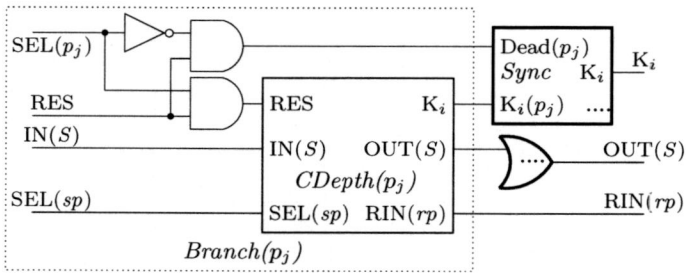

Figure 6.19: Depth circuit for "$p_1 \;||\; \ldots \;||\; p_n$" ($i$ ranges over $\bigcup_{j=0}^{n} K_d(p_j)$)

"...." Extensible gates and sub-circuits accept input from all the branches, even though our figure pictures only the wires coming from branch p_j. For instance, the OUT(S) output of the surface circuit is the disjunction of all the OUT(S) outputs of the branches p_j, $1 \leq j \leq n$. Also, the synchronizer sub-circuit $Sync$, detailed next, has as inputs the Dead(p_j) and K$_i$(p_j) wires output by all the branches.

The behavior of the surface circuit is simple. When the global GO input is 1, all the branches are started by triggering their surface circuits $CSurf(p_j)$. The $Sync$ sub-circuit synchronizes the completion codes of the branches to produce the completion code of the parallel statement.

The depth circuit (Figure 6.19) works similarly except only selected branches of the parallel are resumed. For the unselected branches, the $Sync$ sub-circuit is simply informed of their inactivity.

The synchronizer sub-circuit $Sync$ implements the completion code synchronization operation specified by the semantics of the parallel statement. The synchronizer has two functions:

- Determining the maximum of the completion codes produced by the active branches. To represent the control flow propagation of the COS, 1's are propagated through the synchronizer.

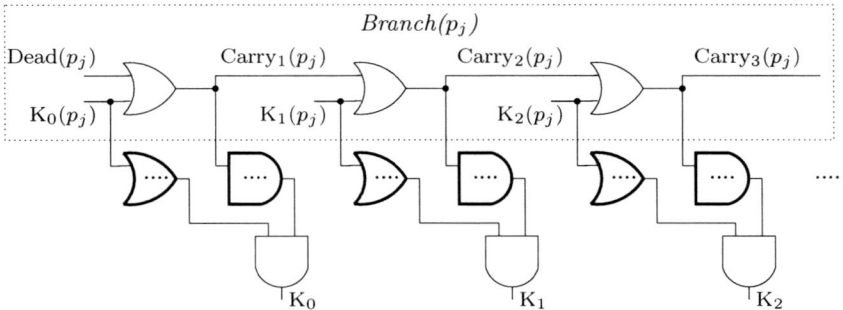

Figure 6.20: The circuit-level parallel synchronizer

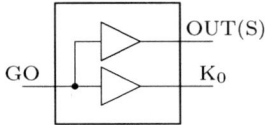

Figure 6.21: Surface circuit for "emit S"

- Pruning the control paths corresponding to completion codes that are constructively determined as unreachable. The computation of the *Can* potential function is represented by propagation of 0's through the synchronizer.

Figure 6.20 shows the synchronizer circuit, which is carefully arranged to allow the parallel, constructive computation of the maximum completion code. When used in the surface translation pattern of the parallel, the Dead(p_j) inputs are set to constant 0, i.e., they are not connected.

Signal declaration and emission

The signal emission statement is instantaneous so it only generates the surface circuit in Figure 6.21, which sets the OUT(S) wire whenever GO is set.

Figure 6.22 shows the surface and depth circuits for signal declaration. Their main characteristic is the feedback from the output OUT(S) of the declared signal back onto the corresponding input IN(S).

Signal test statements

Figures 6.23 and 6.24 shows the surface and depth circuits for both signal and data test statements. Their function is simple: to start the test statement, the GO wire is set to 1, which triggers the *Test* sub-circuit. This sub-circuit determines whether the test expression is true or false, and sets

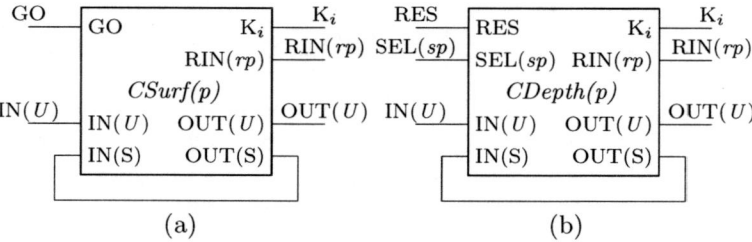

Figure 6.22: Surface (a) and depth (b) circuits for "signal S in p end" (i ranges over $K_s(p)$ in the surface, and over $K_d(p)$ in the depth circuit).

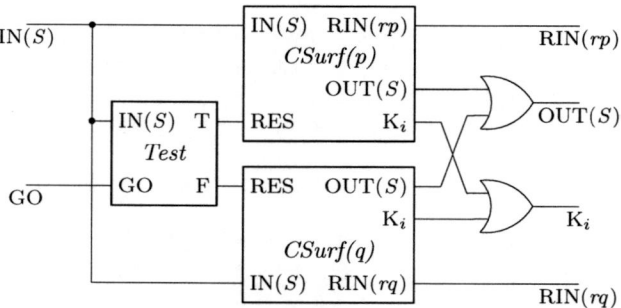

Figure 6.23: Surface circuit for the signal and data tests (i ranges over $K_s(r)$, where r is the test statement).

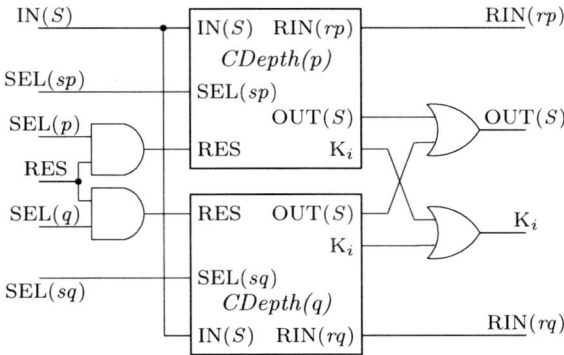

Figure 6.24: Depth circuit for the test statements (i ranges over $K_d(r)$, where r is the test statement).

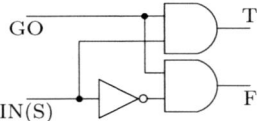

Figure 6.25: Signal test circuit (**present** statement)

the corresponding output to 1, which in turn triggers the surface circuit of the corresponding branch.

The depth circuit resumes the already selected sub-statement. It uses the selection status wires of the two branches to decide which one is selected, and then sets the corresponding RES to 1. Only one branch can be selected at a time.

The only difference between signal and data tests appear in the *Test* sub-circuit. Figure 6.25 shows the version for "**present S then** p **else** q **end**." The next section will give the version for variable tests. More complex test sub-circuits will be presented in Appendix A.1, where we introduce signal expressions.

Suspend

When started, a **suspend** statement behaves like its body, so the associated surface circuit (Figure 6.26) simply instantiates the surface circuit of the body statement.

When resumed, **suspend** performs the suspension test. If the test expression evaluates to 1, control is preempted, the state of the body is frozen by setting the RES wire to 0, and the **suspend** statement completes with code

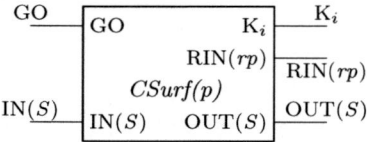

Figure 6.26: Surface circuit for "**suspend** p **when** $expr$" (i ranges over $K_s(p)$)

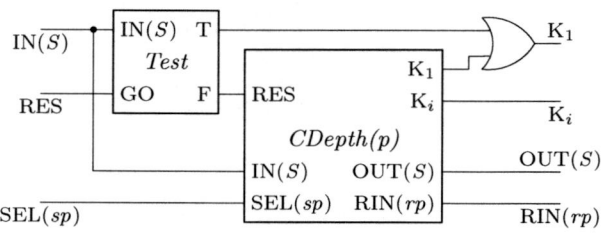

Figure 6.27: Depth circuit for "**suspend** p **when** $expr$" (i ranges over $K_d(\text{suspend } p \text{ when } expr) \setminus \{1\}$).

K_1 (Figure 6.27). We use a *Test* sub-circuit that can also handle the more general case of signal expressions (see Appendix A.1).

Trap exit and handling

The exit statement is instantaneous, so it only generates a surface circuit (Figure 6.28). When triggered, the circuit of **exit T(**i**)** sets the completion code wire K_i to 1.

Figures 6.29 and 6.30 show the surface and depth circuits for the trap declaration statement. Both perform the two essential operations that together form the trap preemption operation: they convert the completion code 2 in normal termination (code 0) and decrement remaining trap codes, and when the body completes with code 2 ($K_2 = 1$), they set all register outputs to 0.

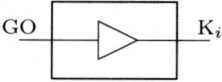

Figure 6.28: Surface circuit for "**exit T(**i**)**"

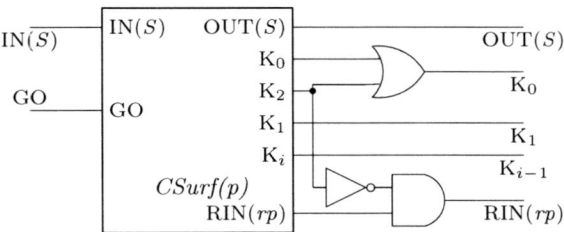

Figure 6.29: Surface circuit for "trap T in p end" (i ranges over $K_s(p) \setminus \{0, 1, 2\}$).

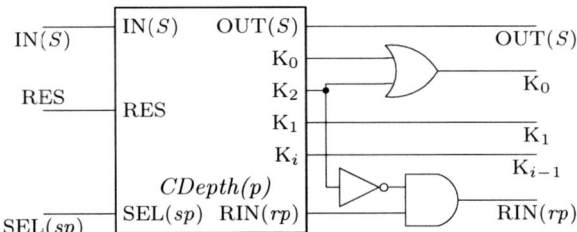

Figure 6.30: Depth circuit for "trap T in p end" (i ranges over $K_d(p) \setminus \{0, 1, 2\}$).

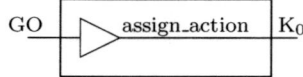

Figure 6.31: Surface circuit for the variable assignment and shared variable emission.

6.3.2 Data-Handling Primitives

Data encoding

The translation encodes the two types of Esterel-level variables onto circuit-level variables in the data store. The encoding is different from that of the COS (Section 5.3), since data dependencies implied by the complex synchronization rules must be represented here using logic gates and dependency arcs.

An unshared variable v is encoded on v_val of type $type_v$. A shared variable s is encoded using two variables (s_val and s_stat) and the causal dependencies defined below.

- s_val, of type $type_s$, holds the actual variable value.

- The Boolean variable s_stat controls the use of the combine function that solves the write-write concurrency. At each evaluation step it is initially 0, and is set to 1 when s has already been assigned to with a statement "$s <= f(\ldots)$".

- Causal dependencies link every assignment "$s <= f(\ldots)$" with every action that reads the value of s. The dependencies are generated at the level of the shared signal declarations.

To represent the data actions, we use small C code fragments that test and modify data variables and return a Boolean result. The return value is only used for test actions.

Note that in typical circuits from Esterel, few wires carry actual data actions. In our formal model, the others carry a trivial action that does not modify the state and returns the value *true* when called.

Data assignment statements

Variable assignment statements are instantaneous, so they only generate the surface circuit in Figure 6.31. The the assign action handles the difference between the shared and sequential variable assignments. According to the previously-defined encoding, the translation of "$s<=f(s_1,\ldots,s_n,v_1,\ldots,v_m)$" generates the following action.

TRANSLATION RULES 127

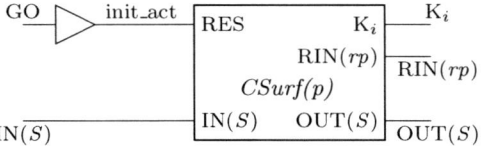

Figure 6.32: Surface circuit for variable declaration statements (i ranges over $K_s(p)$).

```
svar_emit(s):
    if (s_stat==0) {
      s_stat = 1;
      s_val = f(s1_val,...,sn_val,v1_val,...,vm_val);
    } else {
      s_val = c_s(s_val,f(s1_val,...,sn_val,v1_val,...,vm_val));
    }
```

Note that a commutative and associative operator c_s is needed to combine multiple emissions of a shared variable into a single result (Section 2.2).

The action for "$v := f(s_1, \ldots, s_n, v_1, \ldots, v_m)$" is simpler:

```
var_assign(v):
    v_val=f(s1_val,...,sn_val,v1_val,...,vm_val) ;
```

Variable declaration statements

Figure 6.32 shows the surface circuit generated for variable declaration statements. The initializing action is performed just before starting the body statement p. In the case of a declaration of the sequential variable v, the initializing action is

```
var_init(v):
    v_val = init_v;
```

When the statement declares the shared variable s, the initializing action is

```
shared_init(s):
    s_val = init_s;
    s_stat=0;
```

When the shared variable declaration circuit is instantiated, we generate an arc from every svar_emit(s) action to every action that reads s in the surface.

Figure 6.33 gives the depth circuit of variable declaration statements. The reset action is empty for a sequential variable declaration; the action for a declaration of the shared variable s is

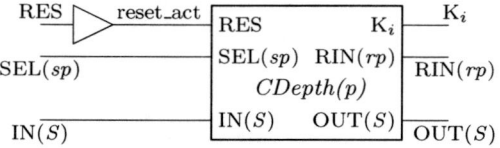

Figure 6.33: Depth circuit for variable declaration statements (i ranges over $K_d(p)$).

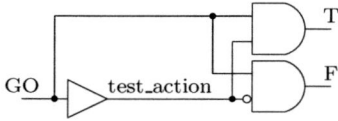

Figure 6.34: Variable test circuit (`if` statement)

```
shared_reset(s):
    s_stat = 0;
```

For shared variables, we also add dependency arcs in the depth just as we do in the surface.

Variable tests

The variable test statement uses the same pattern as the **present** statement; Figure 6.34 shows the test circuit. When translating the primitive variable test statement "`if` v `then` p `else` q `end`," the test action is

```
test_action(v):
    return v_val;
```

More complex test expressions generate more complex circuits.

6.4 Circuit Translation versus COS

In this section we explain how evaluating a circuit with data represents the behavior of the source program according to the COS. Specifically, we will show the start behavior is performed by evaluating the surface circuit and the depth behavior is performed by the depth circuit.

In what follows, P is a Kernel Esterel program of body statement p, I is the input signal set of P, O the output signal set, IS the set of input shared variables, and OS is the set of output shared variables. S is the set of shared variables of p, and V is the set of sequential variables in p. We assume $I \cap O = \emptyset$ and that each variable in the program has a unique name.

Data encoding

The stores used by the COS to express the behaviors of P assign the variables of $OpVar = V \cup \{s_{status}, s_{value} \mid s \in S\}$. Let $OpStores$ be the set of store valuations over $OpVar$, in the encoding of Section 5.3.

The stores built by the circuit translation use a different encoding of the shared variable synchronization. These stores assign the variables of $CVar = V \cup \{\texttt{s\_stat}, \texttt{s\_val} \mid s \in S\}$. Let $CStores$ be the set of store valuations over $CVar$, as defined in Section 6.3.2.

Given $data \in OpStores$, we define its image $[data]$ in $CStores$ by

$$\begin{aligned}
{[data]}(v) &= data(v) \\
{[data]}(\texttt{s\_val}) &= data(s_{value}) \\
{[data]}(\texttt{s\_stat}) &= \begin{cases} 0, & \text{if } data(s_{status}) = old \\ 1, & \text{otherwise} \end{cases}
\end{aligned}$$

for all $v \in V$ and $s \in S$.

Start behavior representation

Start behaviors of the considered program are defined as maximal COS derivations.

$$\bullet p, data \xrightarrow[E]{E',k} \ddot{\overline{p}}, data'$$

We represent this derivation through an evaluation of the surface circuit $CSurf(p)$ associated with the program body.

In the sequel, we assume $\mathcal{W}_S$ is the wire set of $CSurf(p)$ and $\mathcal{W}_D$ is the wire set of $CDepth(p)$. The initial state of the circuit, where the evaluation starts, is defined to correspond to the initial state and input event of the COS derivation.

$$\dot{CSurf(p)}_0 = \langle \dot{\mathcal{W}}_S^0, [data] \rangle$$

where the wire statuses define the initial circuit valuation

$$\dot{\mathcal{W}}_S^0(w) = \begin{cases} E(w) & \text{if } w \in I, \\ 1 & \text{if } w = GO, \text{ and} \\ \bot & \text{otherwise.} \end{cases}$$

Starting in this state, we let the circuit perform a maximal evaluation sequence, and let

$$\dot{CSurf(p)}_1 = \langle \dot{\mathcal{W}}_S^1, data^1 \rangle$$

be its final valuation.

The relation is defined by the following conjecture.

Conjecture 4 *The evaluation sequence assigns a value of 0 or 1 to every circuit wire iff the term $\ddot{\overline{p}}$ contains no control. In this case,*

1. for all $w \in O$ we have $E'(w) = \dot{\mathcal{W}}_S^1(w)$;
2. $k = 0$ iff $\dot{\mathcal{W}}_S^1(K_0) = 1$;
3. $k = 1$ iff $\dot{\mathcal{W}}_S^1(K_1) = 1$;
4. the **pause** statement rp is selected in $\ddot{\overline{p}}$ iff $\dot{\mathcal{W}}_S^1(RIN(rp)) = 1$;
5. for all $v \in V$, $data^1(v) = data'(v)$; and
6. for all $s \in S$, $data^1(\mathtt{s\_val}) = data'(s_{value})$.

Resumption behavior representation

Resumption behaviors of a program are defined as maximal COS derivations.

$$\bullet \widehat{p}, data \xrightarrow[E]{E',k} \ddot{\overline{p}}, data'$$

We represent this derivation through an evaluation of the surface circuit $CDepth(p)$ associated with the program body.

The initial state of the circuit, where the evaluation starts, is defined to correspond to the initial state and input event of the COS derivation.

$$CDepth(p)_0 = \langle \dot{\mathcal{W}}_D^0, [data] \rangle$$

where the wire statuses define an initial circuit valuation

$$\dot{\mathcal{W}}_D^0(w) = \begin{cases} E(w) & \text{if } w \in I, \\ 1 & \text{if } w = \text{RES}, \\ 1 & \text{if } w = \text{SEL}(sp) \text{ and } sp \text{ selected in } \widehat{p}, \\ 0 & \text{if } w = \text{SEL}(sp) \text{ and } sp \text{ not selected in } \widehat{p}, \text{ and} \\ \bot & \text{otherwise.} \end{cases}$$

Starting in this state, we let the circuit perform a maximal evaluation sequence, and let

$$CDepth(p)_1 = \langle \dot{\mathcal{W}}_D^1, data^1 \rangle$$

be its final valuation.

Then the relation is defined by the following conjecture.

Conjecture 5 *The evaluation sequence assigns a value of 0 or 1 to every circuit wire iff the term $\ddot{\overline{p}}$ contains no control. In this case,*

1. for all $w \in O$ we have $E'(w) = \dot{\mathcal{W}}_D^1(w)$;
2. $k = 0$ iff $\dot{\mathcal{W}}_D^1(K_0) = 1$;
3. $k = 1$ iff $\dot{\mathcal{W}}_D^1(K_1) = 1$;
4. the **pause** statement rp is selected in $\ddot{\overline{p}}$ iff $\dot{\mathcal{W}}_D^1(RIN(rp)) = 1$;
5. for all $v \in V$, $data^1(v) = data'(v)$; and
6. for all $s \in S$, $data^1(\mathtt{s\_val}) = data'(s_{value})$.

Sequential behavior

The evaluation of *CSurf(p)* with GO set to 0 sets all the non-input wires (including the outputs) to 0, and triggers no data computation. Similarly, the evaluation of *CDepth(p)* with RES set to 0 sets all outputs to 0 and triggers no data computation. The sequential behavior of the global circuit associated to a program (cf. Figure 6.11) combines the evaluations of the surface, depth, and selection circuits associated to the program. The surface circuit performs the computation of the first instant and the depth circuit computes all the other instants.

Part III

Compiling Esterel

7
Overview

The first two parts of this book presented the Esterel language and its formal semantics. We started with a description of the synchronous/reactive paradigm and an overview of Esterel's syntax. In the second part, we presented the formal semantics of Esterel and described its translation into circuits.

In this part, we explain how to compile the Esterel language by showing how its semantics can be mapped into efficient sequential C code. In this chapter, we give an overview of existing compilation techniques. Details follow in subsequent chapters.

7.1 Compiler Classes

As explained in Section 1.3, all current software implementations of Esterel have the same global structure: an executive repeatedly calls a reaction function that computes the reaction of an instant. Compilers differ in the structure and efficiency of the code they generate for the reaction function.

All current Esterel compilation techniques can be divided into three broad classes based on the intermediate representation they use, which in turn influences the form of the generated code. Figure 7.1 illustrates these classes, which correspond roughly to the three semantics defined in the second part of this book.

Explicit FSM code is generated by symbolically interpreting the program to produce a global, flat, and unstructured (Mealy) finite state machine, which is then encoded in C. This corresponds to explicitly determining all behavioral semantics transitions the program can make in all of its reachable states.

Circuit code is generated by interpreting the Esterel program as a netlist. Software is generated that simulates this netlist, i.e., that computes the value of all wires and registers in an ordered manner at each reaction.

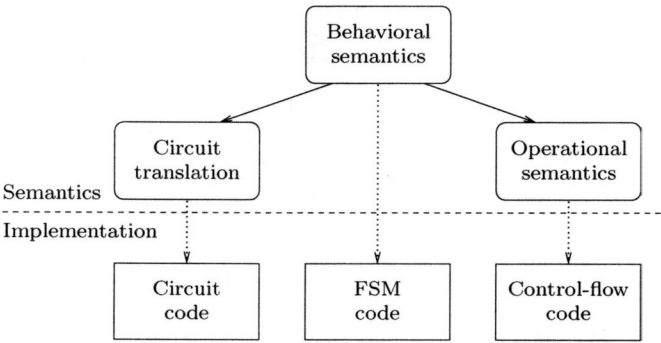

Figure 7.1: Esterel implementation flavors and the corresponding semantic control flow paradigms .

Control-flow code is generated by directly translating the operations and control flow propagation of the COS. The goal is to translate it into a more highly structured model whose semantics more closely match that of the target processor.

We compare the outputs of the various compilation techniques in Sections 9.4 and 10.3.

7.2 A Brief History

The first Esterel compilation techniques (Esterel V2, Esterel V3) generated FSM code. FSM code is theoretically as fast as possible since all state transitions are pre-computed. Unfortunately, this makes the size of the generated code grow rapidly with the specification size (potentially exponentially), which makes the method intractable for real-size applications.

The second generation of Esterel compilers (Esterel V4, Esterel V5) generated circuit code so large programs could be compiled. The use of a digital circuit as an intermediate representation also provided a strong link with existing tools for digital circuit design. The main drawback of circuit code generators is the low speed of the generated code. In particular, the compiled circuit simulators evaluate every circuit gate at each execution instant, meaning the reaction time grows linearly with the specification size rather than the amount of work the program tries to accomplish in each reaction.

The next breakthrough came in the late 1990s, when two research teams [27, 21] independently developed compilers that generated executables that were both small and fast. The new techniques followed from the observation that much of Esterel's COS is based on classical control flow. Thus, it seemed natural to design a compilation technique that relied on direct translation of the sequential operations in the language, making use of the native control-flow constructs of the sequential language (C) into which Esterel is translated. Such a *control-flow code* combines advantages of the FSM and circuit code generators, and can be seen as intermediate between the two.

As in the FSM-based compilers, speed is achieved by efficiently structuring the generated code so that only code for semantically-active parts of the program is traversed. The difference is that the new techniques do not perform the potentially exponential FSM expansion, relying instead on structural information preserved from the initial Esterel specification.

As in the circuit code generators, the translation is largely syntax-driven, and of quasi-linear complexity. The difference is that most of the program structure is preserved and that the intermediate representations are close to the Esterel semantic level, which allows more compact representations that can be easily encoded into efficient software.

In all approaches, control-flow code is comparable in size to, or smaller than the circuit code, while approaching the speed of the explicit FSM code.

Defining a control flow-based approach involves two main difficulties. The first is the choice of an intermediate representation level that has high-level operators that can be easily encoded into efficient software, allows efficient code transformations and optimizations, and allows a clear mapping of the formal Esterel semantics at the intermediate level.

The second problem is to correctly schedule the computation of parallel branches with minimal overhead. The first control flow code generators were based on static scheduling of the reactive operations, meaning that only acyclic Esterel specifications could be handled. One issue is that defining "acyclic" is not simple. We discuss this issue in Section 9.1.

Subsequent research on Esterel compilation has concentrated on defining new representations and associated scheduling techniques.

In the remainder of this chapter, we shall give an overview of existing compilers. The next three chapters provide a detailed description of three significant control flow code generators.

7.3 The INRIA Compiler

The development of the INRIA compiler began in the 1980's and continued for over twenty years by the Esterel team lead by Gérard Berry at the *École des Mines de Paris* and *INRIA Sophia Antipolis*, France. Part of this team, including Berry, left academia in 2001 to join Esterel Technologies, a company that develops and commercializes the Esterel Studio graphical development environment based on Esterel V7. Esterel V7 is a considerable extension of the Esterel V5 dialect. We present it in Appendix D.

Starting in 1984, the Esterel team developed several Esterel compilers, organized on a common platform. As shown in Figure 7.2, all the compilers of the platform share the front-end and the final code generation stages. The front-end takes Esterel sources, translates them into a high-level primitive

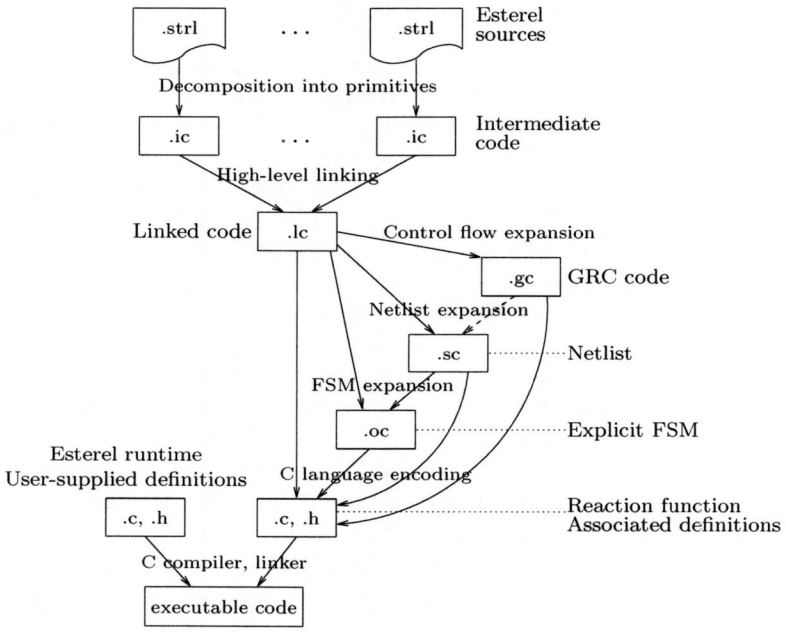

Figure 7.2: The flow of the INRIA compiler. Not pictured are the optimization transformations associated with the GRC, netlist, and explicit FSM representation levels. Boxes list filename extensions.

language (known as intermediate code or IC), and performs sub-module instantiation. The output of this phase is a single linked intermediate code file (LC). The front-end checks the syntactic correctness of the Esterel specification and performs simple semantic checks, not a full constructiveness analysis.

The final code generation stages take the reaction function and the associated definitions (state encoding, etc.) produced by the Esterel compiler and links together with the Esterel runtime and any user-supplied libraries to generate the executable code. The final stages of the compilation process are performed by a compiler for the host language (usually C or C++).

The main work in Esterel compilation is the synthesis of the reaction function. Hence the large number of methods for translating the LC representation into the host language. The first compilers generated explicit FSM code. The FSM expansion was performed first by direct simulation of the program source, then by simulation of a high-level intermediate representation (LC [35] or netlist [6]).

Starting in the early 1990's, compilation was based on the circuit translation [6]. This interpretation, performed at the LC level, is similar to the circuit translation presented in Chapter 6. Examples of the resulting netlist and netlist code are shown in Figures 6.1 and 1.5.

The last component of the INRIA compiler is the GRC-based control-flow code generator that was added in 2002 [56]. We describe it in Chapters 8–9.

7.4 The Synopsys Compiler

The Synopsys compiler [27, 26] was the first to adapt the use of traditional compiling techniques to the compilation of Esterel. It tries to take advantage of the source program structure to produce well-structured code. The example in Figure 7.3 shows how a simple Esterel fragment is translated into code that closely follows the initial program structure. The state is encoded in a hierarchical logarithmic way in the variable s. When control reaches an Esterel "pause" statement, the state changes and the fragment pauses. The C code for the "pause" assigns a new value to the state variable and then branches to the end of the code. The preemption test is activated in instants where the "abort" statement is selected (when s&0x3==3).

However, the translation is not direct. The compiling process starts with the LC representation described earlier. The LC is first translated into a concurrent control-flow graph (CCFG) intermediate representation. In the CCFG, arcs define control and data dependencies, while the nodes define computing actions: assignments, tests, and computation related to parallel branch synchronization. The Figure 7.4 shows a code fragment and the CCFG for it.

The state of the program is encoded on the integer variables s0, s1, and s2. The first encodes the global status of the program; s1 and s2 encode the

```
                        Start: goto L1;
                        Resume:
                          switch (s & 0x3) {
  pause;                  L1:     s=1; goto Join;
  pause;                  case 1: s=2; goto Join;
  abort                   case 2: goto L2;
                          case 3: if(!A)
                            switch (s>>2 & 0x7) {
    pause;                  L2:     s = 3 | 0<<2; goto Join;
    pause;                  case 0: s = 3 | 1<<2; goto Join;
    pause;                  case 1: s = 3 | 2<<2; goto Join;
    pause;                  case 2: s = 3 | 3<<2; goto Join;
    pause                   case 3: s = 3 | 4<<2; goto Join;
                            case 4: ;
  when A                  }
                          s=0; goto Join;
                          case 0: ;
                          }
                        Join:
```

Figure 7.3: Esterel fragment and part of its Synopsys translation into C. After Edwards [26].

status of the two parallel branches. The variable s0 is 2 when the program starts and 1 otherwise; s1 is 0 when the first parallel branch is not selected (terminated), 1 when the first "pause" is selected, and 2 when the second "pause" is selected.

The solid arc in Figure 7.4 indicate flow. The dashed arc is a signal dependency between the emission and the test of S. The code begins by decoding the state through a cascade of conditionals that test both state variables and signals (for, e.g., abort statements). The next instant's state is set by assigning to state variables.

Once the CCFG constructed, it is statically scheduled into sequential C. While it only accepts acyclic intermediate specifications, the scheduling algorithm generates high-quality code by exploiting the exclusivity of control-flow paths. This limits the overhead of context-switching code so the result is compact.

7.5 The Saxo-RT Compiler

The central challenge in compiling Esterel for a single-threaded processor (as opposed to hardware) is the need to interleave the execution of multiple threads of control within a single instant. The Saxo-RT compiler, developed by a group at France Telecom R&D [14, 21, 72], uses a technique based

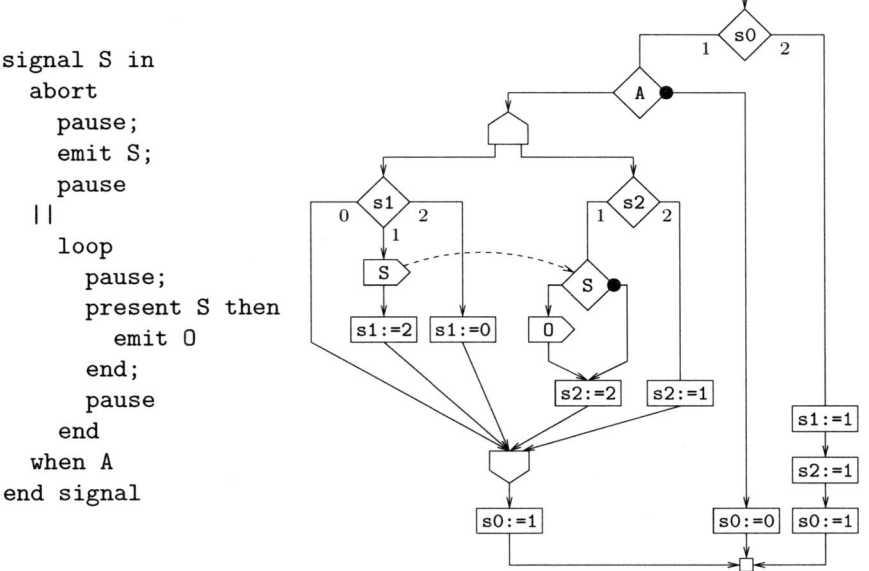

Figure 7.4: A small program and the concurrent control flow graph (CCFG) built by the Synopsys compiler for it.

roughly on the behavior of discrete event simulators (French [33] describes the general architecture of such simulators along with a way to accelerate them).

Unlike the Synopsys compilation technique, which is built around a classical CCFG representation, the Saxo-RT compiler uses an event graph (EG) representation that allows instructions to be scheduled and removed from the schedule in both the current cycle and the next. EG nodes represent small segments of the Esterel program that can execute atomically (i.e., do not cross a **pause** or signal test). The arcs of the EG represent four types of control dependence. The compiler generates code by ordering the nodes according to control and data dependencies and generating a small C function for each node. The reaction function consists of a hard-coded scheduler that steps through the functions and call each if it is currently active.

Control dependence arcs come in four flavors: enabling and disabling for the current and next cycle. Enabling in the current cycle is the simplest: in Figure 7.6, if signal I is present and node f3 is active (runs), the **weak abort** and **sustain R** instructions should run in the same cycle (**sustain R** has been transformed into **emit R** with a self-loop). The "enable current" arcs from f3 to f7 and f4 indicate this.

"Enable next" arcs implement the behavior of instructions such as **pause** and **await**, which wait a cycle before proceeding, by activating their targets

```
signal R, A in
  every S do
    await I;
    weak abort  sustain R  when immediate A;
    emit O
  ||
    loop
      pause; pause;
      present R then emit A end present
    end loop
  end every
end signal
```

Figure 7.5: A simple Esterel module modeling a shared resource. The first thread generates internal requests (R) in response to external requests (I), and the second thread acknowledges them (A) in alternate cycles. The S input resets both threads.

in the next cycle. For example, when the outer "every S" statement runs, it starts the two threads that begin with "await I" and pause. The "enable next" arcs leading from f0 to f2 and f3 activate these statements in the next cycle. The f0 node also uses such an arc to schedule itself in the next cycle, which ensures signal S is checked in the next cycle.

By default, active nodes whose conditions are not met (e.g., f3 is active but signal I is not present) remain active in the next cycle. Thus, when a statement does run, it usually disables itself. Self-loops with disable arcs, such as those on f5, f2, and f6, accomplish this.

Preemption instructions also use disable arcs. For example, when f7 is active and signal A is present, f7 preempts its body (which contains f1 and f4) by disabling them in both the current and next cycles. Node f0, which preempts most of the program, has many such disable arcs.

The compiler encodes the event queue as a bit vector. Each node is assigned a bit in an integer variable, using multiple words if the number of nodes exceeds the processor's word size. Logical operations on these variables add and remove nodes from the event queue.

Nodes are ordered according to control and data dependencies to generate the final linear schedule. For example, nodes f1 and f4 both emit the R signal and node f6 checks it, thus f6 appears later in the schedule than f1 or f4. Control dependencies also impose ordering constraints. Because f7 is a weak abort, which only preempts its body (f1 and f4) after it has had a chance to execute for the cycle, f7 appears after f1 and f4.

The Saxo-RT compiler rejects program whose nodes have no linear order. We discuss different compilers' notions of acyclicity in Section 9.1.

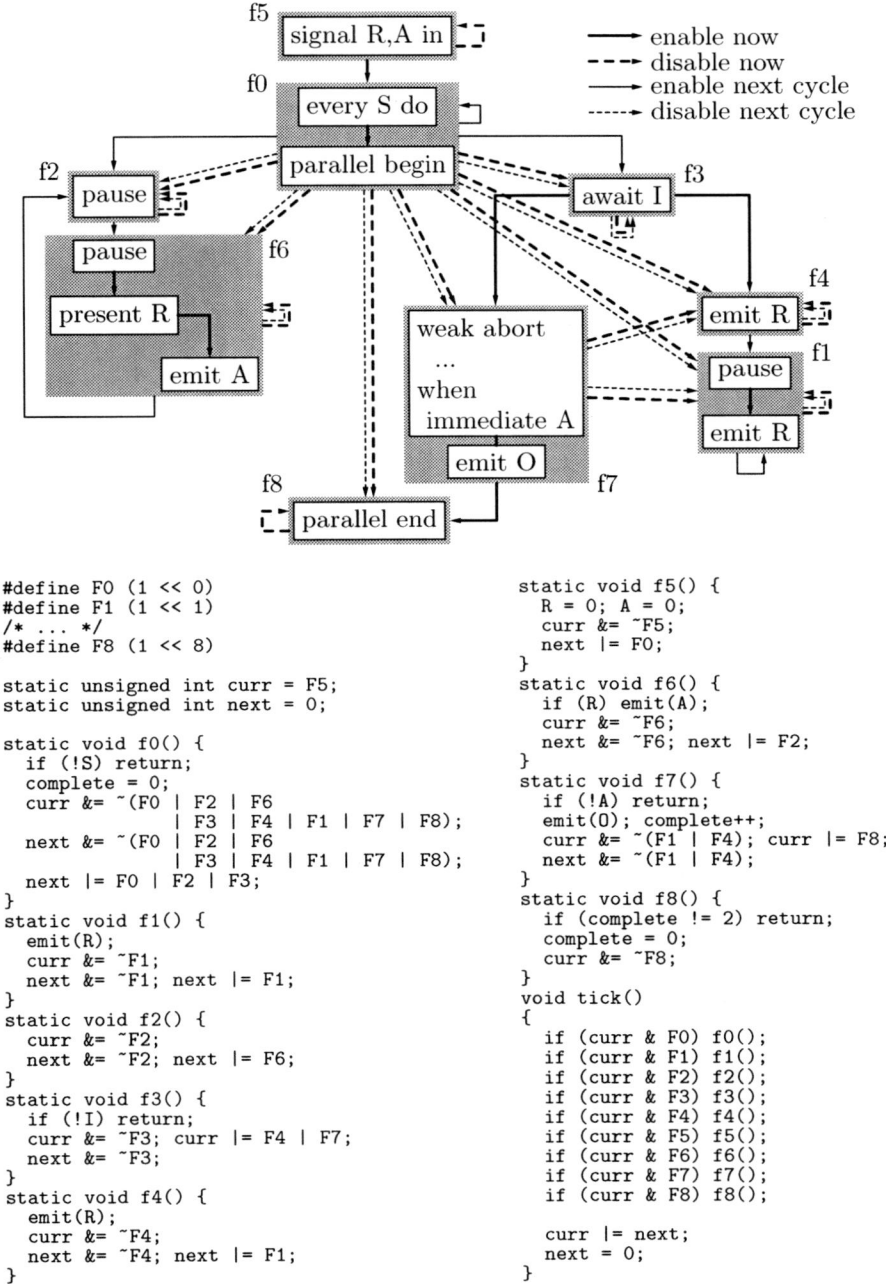

```
#define F0 (1 << 0)
#define F1 (1 << 1)
/* ... */
#define F8 (1 << 8)

static unsigned int curr = F5;
static unsigned int next = 0;

static void f0() {
  if (!S) return;
  complete = 0;
  curr &= ~(F0 | F2 | F6
              | F3 | F4 | F1 | F7 | F8);
  next &= ~(F0 | F2 | F6
              | F3 | F4 | F1 | F7 | F8);
  next |= F0 | F2 | F3;
}
static void f1() {
  emit(R);
  curr &= ~F1;
  next &= ~F1; next |= F1;
}
static void f2() {
  curr &= ~F2;
  next &= ~F2; next |= F6;
}
static void f3() {
  if (!I) return;
  curr &= ~F3; curr |= F4 | F7;
  next &= ~F3;
}
static void f4() {
  emit(R);
  curr &= ~F4;
  next &= ~F4; next |= F1;
}
static void f5() {
  R = 0; A = 0;
  curr &= ~F5;
  next |= F0;
}
static void f6() {
  if (R) emit(A);
  curr &= ~F6;
  next &= ~F6; next |= F2;
}
static void f7() {
  if (!A) return;
  emit(O); complete++;
  curr &= ~(F1 | F4); curr |= F8;
  next &= ~(F1 | F4);
}
static void f8() {
  if (complete != 2) return;
  complete = 0;
  curr &= ~F8;
}
void tick()
{
  if (curr & F0) f0();
  if (curr & F1) f1();
  if (curr & F2) f2();
  if (curr & F3) f3();
  if (curr & F4) f4();
  if (curr & F5) f5();
  if (curr & F6) f6();
  if (curr & F7) f7();
  if (curr & F8) f8();

  curr |= next;
  next = 0;
}
```

Figure 7.6: (top) The event graph the Saxo-RT Esterel compiler generates for the program in Figure 7.5. Each gray area is a node that becomes a single function in the generated code (bottom). Control and communication dependencies between these groups dictates the order in which they appear in the reaction function, named `tick()`.

7.6 The Columbia Esterel Compiler

The Columbia Esterel Compiler (CEC) is a compiler system developed at Columbia University by the research group of Stephen A. Edwards. Starting from an effort to extend the Synopsys compiler, the compiler system now includes a number of novel techniques, which we present in Chapter 10.

8
The GRC Intermediate Format

Digital circuits (i.e., gate-level schematics) form a natural model for reasoning about the reactive behavior of Esterel programs (but not their structure). Perhaps the most direct definition of Esterel's semantics is given through translation into constructive circuits. Also, among the sequential code generation schemes available today, the one based on circuit simulation is the most used. Unfortunately, three characteristics make circuit netlists a bad starting point for the generation of fast C code for a variety of reasons.

The fine grain of the representation is one problem. The C code that simulates a netlist uses large numbers of Boolean "wire" variables, computed from one another using lots of low-level Boolean operators.

The lack of structure is another. The constructive simulation of a circuit involves the evaluation of every gate at each execution instant.

Finally, the state encoding, based on registers, requires each memory component to be re-initialized at each execution instant.

To preserve some of the desirable properties of circuits in a formalism that is more adapted to sequential code generation, the control-flow code generators have defined new representations with larger operators such as tests and procedure calls.

Here, we present the GRC ("GRaph Code") format and the associated code optimization and code generation techniques used in the control-flow code generator of the INRIA compiler. GRC can be seen as more abstract than similar formalisms used in the Synopsys and Saxo-RT compilers. Fewer encoding choices are made at the GRC level, allowing more optimizations to be performed prior to software generation. Finally, the GRC also preserves a strong link with the circuits so it clearly reflects the constructive semantics of Esterel in an operational fashion.

Variants of the GRC format are used in the Columbia Esterel compiler, presented in Chapter 10, and in the Esterel V7 compiler used in the Esterel Studio graphical development environment presented in Appendix D.

A specification in the GRC format defines two things: a control/data flowgraph and a hierarchical state representation called the selection tree. These two are connected through primitives that indicate how the flowgraph tests and updates the state representation. A set of variables represents user-defined variables. Additional variables hold the hierarchical state representation, i.e., the selection status of every statement in the program.

We can see GRC specifications as an intermediate step of the circuit translation where the hierarchical state has not yet been flattened and encoded using latches and logic gates, and behavioral constructs like tests or parallel synchronization have not yet been expanded into gates.

The operators in GRC are similar to those in the intermediate representations of the Synopsys [27] and Saxo-RT [21] compilers. Yet when we interpret a GRC specification using constructive semantics we obtain a model capable of representing the entire class of Esterel programs. This model is valuable as it allows us to reason at the level of GRC. It makes it possible to define and prove properties such as the correctness of code optimizations.

The translation of Esterel into GRC is structural, pattern-based, and very similar to the circuit translation of Chapter 6. The resulting GRC specifications can be optimized using three types of techniques:

- general constructive circuit optimizations that still apply, such as sweeping and constant value propagation;

- general control-flow optimizations, such as the removal of dependencies between branches of a test; and

- Esterel-specific optimizations, based on static analysis of the hierarchical state representation.

In the end, C code is generated by scheduling the computation of the flowgraph nodes.

This chapter presents the GRC format in two parts. The first defines the format, the structural translation of Esterel into GRC, and the formal simulation semantics of GRC. The second part presents some GRC optimization techniques.

8.1 Definition and Intuitive Semantics

GRC is a textual format that also has a graphical representation. We use the graphical form to explain, but present some textual examples in Section 8.1.3.

8.1.1 The Hierarchical State Representation

The state representation of a GRC specification is based on a tree structure called the selection tree, composed of typed selection nodes. More precisely, the selection tree is a parallel/exclusive abstraction of the abstract syntax tree of the Esterel program. The selection tree has the same structure as the

DEFINITION AND INTUITIVE SEMANTICS 147

```
module MainExample:
input I, J, KILL, SUSP; output O; % interface declarations
suspend
  trap T in % performs the preemption
    signal END in
      loop % basic computation loop
        await I ; emit O ; await J ; emit END
      end
    ||
      % preemption protocol, triggered by KILL
      await KILL ; await END ; exit T
    end
  end;
when SUSP % suspend signal
end module
```

Figure 8.1: A simple Esterel program modeling a cyclic computation (like a communication protocol) that can be interrupted between cycles and suspended.

state representation of the circuit translation (the selection circuit, defined in Section 6.2), but stores more information—most notably exclusions—needed to generate efficient software.

Pause statements generate leaf nodes, sequential composition and tests—**present** or **if**—produce exclusive nodes, while parallel constructs are preserved as such. Composite statements with only one argument (e.g., loop, suspend) generate reference selection nodes with only one child. Instantaneous statements, such as **nothing** and **emit**, are ignored because they do not hold control between execution instants.

We use MainExample as a running example. Its source is in Figure 8.1; its selection tree is Figure 8.2.

Each label indicates the statement corresponding to the node. Squares represent **pause** statements. Exclusive nodes are marked with #; the unique parallel node is marked with ||. The indices are from enumerating the nodes, a process we present on page 154. We will define the meaning of the nodes of indices 0 and 1 in Section 8.2. Selection nodes may later be tagged for optimization and code generation purposes (see Section 8.1.3).

Each selection node implicitly defines a Boolean flag that records the current selection status of its associated Esterel statement. By extension, the flag is called the selection status of the associated selection node, hence the designation "selection tree." The GRC-level selection status bits are checked and modified by state access primitives triggered from the flow graph. As for an Esterel statement, we say that a selection node is selected if its selection

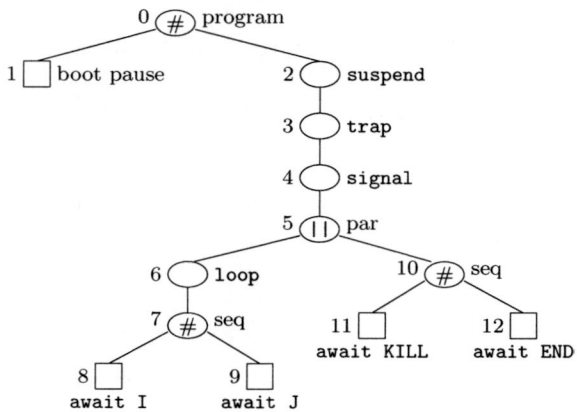

Figure 8.2: The selection tree of MainExample

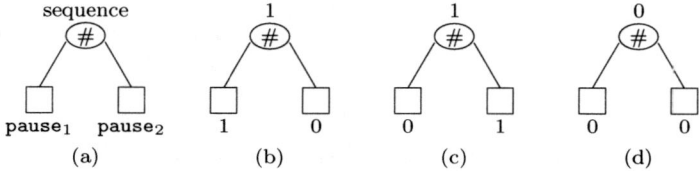

Figure 8.3: Selection tree and selection flags for a small example

status is true. Otherwise we say it is not selected.

To see how the encoding works, consider the statement "pause ; pause." Its selection tree is pictured in Figure 8.3(a). The selection status bits associated to nodes to represent the (macrostep) states "$\widehat{\text{pause}}$; pause" and "pause ; $\widehat{\text{pause}}$" are pictured in Figure 8.3(b) and (c) respectively. The remaining valid valuation of the selection status bits, corresponding to the statement being not selected, is pictured in Figure 8.3(d). Note that not every valuation of the selection status bits represents a program state. We say a valuation of the selection status bits is consistent if a selection node is selected *iff* it is a selected pause node or it has a selected child and an exclusive node never has more than one selected child.

The computation of a reaction always starts with the selection status bits having a consistent valuation. The status bits may be changed during computation, but are left consistent at the end of the reaction. Consistency is essential to the following definitions.

The GRC flowgraph (defined later) reads and updates the selection flags using five state access primitives: *enter*, *exit*, *test*, *switch*, and *sync*. Each primitive is associated with a selection node. A *switch* is always associated

DEFINITION AND INTUITIVE SEMANTICS   149

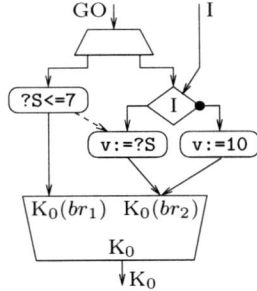

Figure 8.4: Simple GRC flowgraph

with an exclusive node. Similarly, a *sync* always has a parallel node.

The state is decoded by the control-flow graph through calls to the state decoding primitives *test* and *switch*. The first one simply returns the status of its argument. *Switch* returns the unique selected child of the argument. The control-flow graph decodes the state hierarchically: the status of the root node is tested first. If it is selected, the status bits of its children are tested, etc. The decoding stops at unselected nodes and at leaves.

The control-flow graph updates the state using the state update primitives *enter* and *exit*. A call to *enter* sets the the selection status of its argument to true. Executing *exit* sets to false the selection status bits of all the nodes in the sub-tree rooted in the argument node. For instance, to change the state of "pause ; pause" from "$\widehat{\text{pause}}$; pause" to "pause ; $\widehat{\text{pause}}$," one has to execute "$exit(\text{pause}_1)$" and "$enter(\text{pause}_2)$."

The state update is performed by the control-flow graph only after state decoding. Thus, state decoding primitives always check the consistent initial valuation of the selection status bits, before they can be modified by state update primitives.

The *sync* primitive decides whether a parallel statement terminates or pauses in an instant. A parallel terminates when all its branches terminate. The result of *sync* is true if at least one child of the parallel argument selection node is still selected and false otherwise. I.e., if the result is false the parallel does not pause (it either terminates or exits a trap).

8.1.2 The Control/Data Flowgraph

The flowgraph is defined by connecting typed computation nodes with arcs that define the static causal dependencies. The flowgraph in Figure 8.4 corresponds to

```
emit S(7) || present I then v:=?S else v:=10 end
```

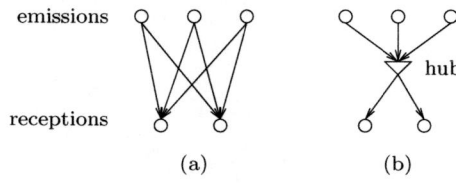

Figure 8.5: Possible representations of signal dependencies. Our choice (a) is to connect every emission with every reception. Another approach (b) is to introduce hub nodes, which minimizes the number of dependencies but can affect optimization.

The six types of computation nodes are **Tick, Test, Join, Call, Switch**, and **Sync**. An arc is either a control arc or a data arc. We define each type of node and arc below; we define the correspondence between Esterel statements and GRC nodes in Section 8.1.3.

Each type of node has a specific interface that consists of *named input and output ports*. A node gathers the control information needed for its computation through its input ports. Control input ports receive the incoming control flow (the bullets in our COS) and signal input ports, present only on test nodes, are used to collect the status bits of the signals involved in the test expression.

A node is executed in an instant if it receives control through at least one input port. In this case, the node will activate exactly one of its output ports after the completion of its execution and based on its result. The output ports are responsible for signal emission and for passing control to other nodes.

Output ports are connected to input ports through control arcs. Each control arc connects one output port to one input port, but each port may have zero or more incident arcs. When activated, an output port activates all its outgoing control arcs. If the destination of an activated arc is a control input, the destination port receives control. If the destination is a signal input port (of a test node), then the signal is considered present in the evaluation of the test. Signal dependencies are control arcs that correspond to signal emission.

The number of signal dependencies is potentially large: one for every emission-test pair for each signal (Figure 8.5(a)). In practice, the number of dependencies remains reasonable for our largest applications, but signal dependencies can be factored by introducing hub objects connected to all emissions and receptions of a signal. This approach is shown in Figure 8.5(b) and has been adopted in the Esterel V7 compiler. We adopt the non-factored form because it allows individual dependencies to be removed one by one.

A node may carry data access operations called actions. Data arcs, like the data dependencies in circuits with data, link these actions to impose an evaluation order that ensures the correct computation of the data variables.

An action can be the source or destination of one or more data dependency arcs.

Below, we list the types of computation nodes and their interfaces.

1. A unique **Tick** node marks the control entry point in each specification. It bears no input ports and a single output port called *CONT*. This triggers the execution of the specification at each computation instant. We draw **Tick** nodes as

   ▽CONT

2. A **Test** node has an input port named *GO*—the test trigger—and two output ports—*THEN* and *ELSE*. Exactly one of these receives control after test executes. The test expression may involve signal status bits or actions—user data tests or selection *test* primitives. For every signal in the test expression, the **Test** node bears a supplementary signal input port.

   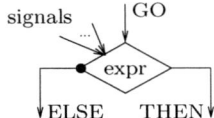

3. **Join** nodes reassemble exclusive control flows (e.g., the branches of a test). They have a single input port, called *GO*, where all incoming control arcs link, and one output port, called *CONT*. When triggered by one of the incoming arcs, the **Join** node instantly passes control to the *CONT* port.

   ▽GO
   ▽CONT

4. **Call** nodes perform data updates. A **Call** node has a single input port "*GO*" and one output port "*CONT*." The data update operation performed by the **Call** is called an action. This action can be a user data operation or a state update primitive. It is executed between when the node receives control and passes control to the output port *CONT*.

5. **Switch** nodes perform a choice associated with exclusive selection-tree nodes. Intuitively, they pass control to the selected branch of a sequence

or test statement. The *switch* state decoding primitive is triggered through the unique *GO* input port. It corresponds to a particular selection node. For every child x of the selection node—for every possible choice—the **Switch** has one output port called $[x]$, through which control is passed. For space considerations, we abbreviate in the graphical representation *switch(node)* with *node*.

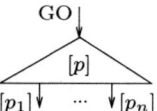

6. **Sync** nodes implement the complex parallel branch synchronization operation defined in Section 3.3.4. They collect the selection status and the completion codes of the parallel branches through input ports and compute the completion status of the parallel statement.

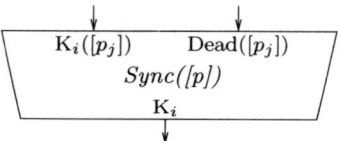

Every **Sync** node is associated with a parallel selection node, which corresponds to an Esterel parallel statement. The children of the parallel node correspond to the branches of this parallel statement. The input ports of the synchronizer gather completion codes for each branch using the selection children as indices. If x is a child of the parallel selection node, then the input port named Dead(br) collects the negated selection status of the associated parallel branch. When it is set to 1 the statement produces no completion code. The input port named $K_i(br)$ collects completion levels of i for the associated branch.

Once the **Sync** computes the completion code of the parallel, it passes control to the relevant handler through output completion code ports K_i. There is one output port per possible completion code. A **Sync** node may also carry a *sync* state access primitive that is used in the code generation process. The computation of the *sync* primitive is redundant in a **Sync** node, but it sometimes allows a better encoding of the synchronization in sequential code.

In addition to the previous nodes, which have textual counterparts, the graphical representation also uses a **Fork** construct to represent places where control forks at the beginning of parallel branches.

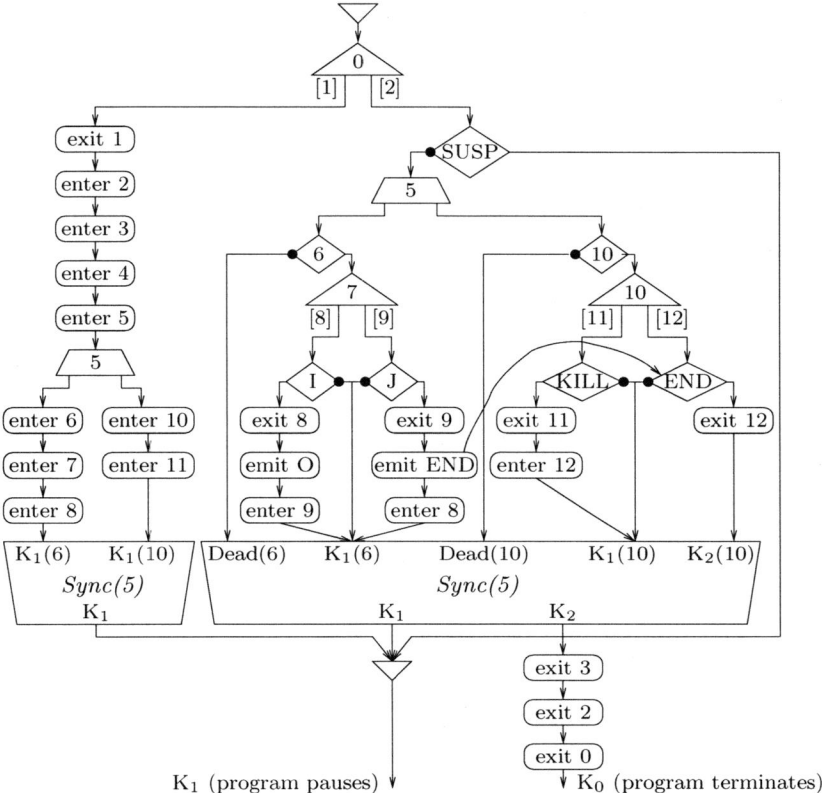

Figure 8.6: The flowgraph of MainExample

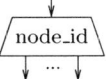

Control forks are only used in the graphical representation. In textual GRC, the node is replaced with a list of successor ports placed everywhere the control fork node needs to be triggered.

Not every specification formed with these elements is meaningful. The Esterel to GRC translation presented in the next section ensures certain well-formedness properties. Essential to code generation, these properties are assumed and preserved by the simplification algorithms. First, the state access primitives must always be called by the flowgraph in a "consistent" order. In particular, state flags are decoded before they are modified within an instant. Furthermore, the graph must be acyclic after removing signal dependencies. In the sequel, we only consider well-formed graphs obtained from

154   THE GRC INTERMEDIATE FORMAT

Esterel programs. Figure 8.6 shows a well-formed graph for `MainExample`. The arguments of the selection primitives are pointers to nodes in the selection tree of Figure 8.2. Most of the port names have been omitted for space reasons. Section 8.1.3 gives fragments of the textual representation of the graph.

8.1.3   Implementation Issues

The GRC format follows the general structuring patterns used in all the intermediate formats developed by the Esterel group [32]. A GRC specification is a text file organized as a sequence of tables. Each table defines all objects of a given type and refers to other objects using integer indices. Most tables are common to all Esterel intermediate formats. Also called data tables, they define data of general use: module instantiation hierarchy, signals and variables, functions and procedures, etc. Each format also defines its own table formats; GRC defines three: the selection node table, the (action) call table, and the (control-flow graph) node table.

Selection tree representation

The selection node table (also called *selnode* table) defines the selection tree. The textual representation of the selection tree of `MainExample` is

```
selnodes: 13
sel:0     exclusive: ( sel:1 , sel:2 )
sel:1     boot: pause:
sel:2     ref: ( sel:3 )
sel:3     ref: ( sel:4 )
sel:4     ref: ( sel:5 )
sel:5     parallel: ( sel:6 , sel:10 )
sel:6     ref: ( sel:7 )
sel:7     exclusive: ( sel:8 , sel:9 )
sel:8     pause:
sel:9     pause:
sel:10    exclusive: ( sel:11 , sel:12 )
sel:11    pause:
sel:12    pause:
end:
```

The table header gives the type of the table (`selnodes:`) and its size. The selection nodes are indexed by numeric selection indices. The node of index `sel:0` is always the root of the selection tree. Keywords indicate the type of each node; children are listed in parentheses.

We defined the selection node types `pause:`, `exclusive:`, and `parallel:` earlier in this chapter. Nodes of type `ref:` have only one non-void child. Such nodes correspond to composed statements like `loop` or `signal` that only have

DEFINITION AND INTUITIVE SEMANTICS 155

one direct sub-statement and statements with multiple branches where only one branch has a non-void selection node.

Nodes of type void: correspond to statements that cannot be selected. They are preserved only as reminders of the initial program structure. Section 8.4 explains them in more detail.

Other tags, such as nonterm:, can be added by the analysis and optimization algorithms. These will be described in Section 8.4.

Flowgraph representation

Control and data dependencies are represented separately in the node and call tables. This separation is motivated by the fundamental difference between signal and data causality in Estereland facilitates analysis and code generation. The call table represents data accesses and data causality; the node table represents constructive control flow and signal dependencies.

The call table lists all the data accesses of the reaction function and all data dependencies among them. Data accesses—procedure and function calls, signal initializations, signal emissions, and state accesses—are dubbed "action calls," hence the name "call table." Action calls are activated from the control-flow graph. The following fragment belongs to the GRC representation of the wristwatch example (part of the INRIA Esterel V5_92 distribution).

```
calls: 825
...
call:7      init:sig:4
...
call:169    swap:sig:2
...
call:383    exit:sel:63
call:384    sync:sel:6
call:385    test:sel:64
call:386    switch:sel:65
call:387    exit:sel:66
call:388    call:act:46
...
call:398    demit:sig:25(incarn:1) act:42
call:399    dumb:(call:609,call:677,call:695,
                  call:698,call:730,call:733,call:750)
call:400    pemit:sig:15(incarn:1)
call:401    enter:sel:68
...
end:
```

Each call has an index and may have a list of data dependencies: indices of calls whose execution must wait until the current call has either been executed or ruled out. For example, the call of index call:399 must be always executed or invalidated before the calls of indices 609, 677, 695, 698, 730, 733, and 750.

There are several types of calls:

- Selection calls execute one of the five state test or update primitives on a specific selection node (represented by its index).

- `call:` entries execute a user-defined action.

- `dumb:` entries do not perform computation but can define data dependencies. Such entries are generated in the automated translation process.

- The signal (re-)initialization calls `init:` and `swap:` mark the beginning of a signal scope at statement start and resume respectively. They are employed in the optimized encoding of signals for sequential code generation.

- The valued signal emission calls `demit:` represent the update of shared variables. Pure signal emission calls `pemit:` represent the action associated with the emission of pure output signals.

The node table represents the flowgraph nodes—the control flow that drives the calls. Below is a fragment of the node table of `MainExample`.

```
nodes: 45
node:0    Tick:(cont: go:@node:1)
node:1    SelSwitch:call:0 (sel:1 go:@node:2)
                         (sel:2 go:@node:14)
node:2    Call:call:1(cont: go:@node:3)
node:3    Call:call:2(cont: go:@node:4)
...
node:19   Test:( sig:3(incarn:1) ) (then: go:@node:20)
                                   (else: go:@node:21)
...
node:21   Call:call:19(cont: go:@node:22)
node:22   Call:call:20(cont: go:@node:23 go:@node:33)
...
node:29   Test:( sig:1(incarn:1) ) (then: go:@node:30)
                                   (else: go:@node:32)
node:30   Call:call:26(cont: go:@node:31
                             sig:6(incarn:1)@node:38)
node:31   Call:call:27(cont: k:1(sel:6)@node:40)
...
node:33   Test:( call:29 ) (then: go:@node:34)
                           (else: dead:(sel:10)@node:40)
node:34   SelSwitch:call:30 (sel:11 go:@node:35)
```

```
                         (sel:12 go:@node:38)
node:35   Test:( sig:2(incarn:1) ) (then: go:@node:36)
                                  (else: k:1(sel:10)@node:40)
...
node:38   Test:( sig:6(incarn:1) ) (then: go:@node:39)
                                  (else: go:@node:37)
node:39   Call:call:33(cont: k:2(sel:10)@node:40)
node:40   Sync:sel:5 call:34 (k:1 ) (k:2 go:@node:41)
node:41   Call:call:35(cont: go:@node:42)
...
end:
```

The textual definition of a node consists of its index, its type and parameters, and a list of named output ports. **Tick** and **Join** nodes do not have parameters. A **Call** node takes a call index parameter. A **Switch** node takes the call index that points to the appropriate *switch* selection primitive as a parameter. The **Test** nodes have a test expression parameter, which is composed from atoms—signal port names and call indices—with the operators not:, and: and or:. A **Sync** node takes the selection index pointing to the associated parallel node as a parameter. It may also take the call index of a *sync* primitive as a parameter. There are no explicit **Fork** nodes in the textual representation; they are implicit in places where control forks, such as the cont: output port of the **Call** of node:22.

Each output port definition consists of the port name and a list of destination port references. For example, the output port cont of the **Call** node 30 is linked to two input ports: the control input port go of the node 31, and the signal port sig:6(incarn:1) of the node 38. This cont output port corresponds to the "emit END" statement of the MainExample program, which gives control in sequence to the parallel synchronizer (code K_1, represented by flowgraph node 31) and emits the signal END which is read by the test of the second parallel branch (node 38).

A node does not list its input ports; this information can be obtained by a forward traversal of the flowgraph. For example, the **Test** node 38 has the input ports go and sig:6(incarn:1).

The definition of a signal port, e.g., sig:6(incarn:1) includes the signal name, as a reference to the signal table (sig:6, which points to the definition of signal END), and the incarnation index incarn:1. The incarnation index is useful mostly for tracing. It identifies the copy number of each signal produced during the translation process. We describe this form of reincarnation, similar to the circuit-level reincarnation mechanism presented in Section 6.3, in the next section. In our example, sig:6(incarn:1) is the depth instance of the END signal.

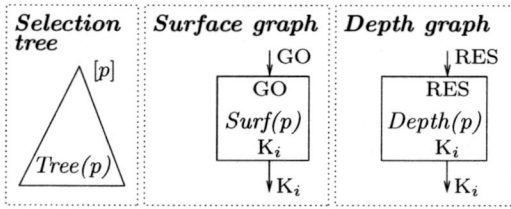

Figure 8.7: The Esterel to GRC translation interface

8.2 Esterel to GRC Translation

In this section, we show how Esterel programs are translated into GRC. The translation is fundamentally based on the COS: the selection tags of the COS are organized to form the selection tree, and the GRC control/data flowgraph represents all possible paths that program counters can take during the execution of an instant (more precisely, a syntactically-determined super-set).

The control flow unfolding that transforms the Esterel source in the intermediate flowgraph closely follows the circuit translation of Chapter 6. This results in important topological similarities that prove useful during analysis and code generation. In particular, these similarities will allow us to relate the notion of cyclicity defined at the circuit level with that at the GRC level (Section 9.1).

8.2.1 Translation Principles

The translation is structural and very similar to the circuit translation of Chapter 6. With every statement p it associates the selection (sub-)tree $Tree(p)$; the *surface flowgraph* $Surf(p)$, representing the start behavior of our statement; and the *depth flowgraph* $Depth(p)$, which represents the behavior of p at instants where it is restarted.

As in the circuit translation, instantaneous statements have no depth code. We associate a selection node with every statement except the simple instantaneous ones (exit, emit, and variable assignment). We denote the selection node associated with a statement p by $[p]$.

To ensure correct signal links, for every signal used in p but declared outside its scope, we maintain the set of input ports reading it and the set of output ports corresponding to its emission. This must be done separately for the surface and depth graphs. For a given signal S, we call the four signal link sets $SurfIn(S)(p)$, $SurfOut(S)(p)$, $DepthIn(S)(p)$, and $DepthOut(S)(p)$. By convention, the signal link sets are not defined for a statement p and signal S when p does not read or write S.

Figure 8.7 shows the interface of the generated graphs. The selection tree

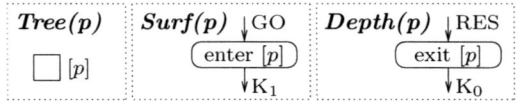

Figure 8.8: The translation of p="**pause**"

Tree(p) is rooted at [p]. The input and output arcs of *Surf(p)* and *Depth(p)* represent the input and output ports, which form the control interface. Giving control to the input port labeled GO corresponds to starting the statement; giving control to RES resumes it. The activation of the output port labeled K_i corresponds to the completion of the computation of an instant with code i. As in the circuit translation, i ranges over the set of potential completion codes of p, as over-approximated by $K_s(p)$ and $K_d(p)$ in the surface and the depth graphs, respectively (cf. Sections 4.6 and 6.3).

The encoding of Esterel-level data onto the stores is identical to that used in the circuit translation (Section 6.3.2). Netlist-level actions are called in the GRC by **Call** and **Test** nodes.

Keeping the surface and depth flowgraphs separated involves some code duplication when Esterel statements are shared between the start and resume behaviors. The duplication follows the same rules as our circuit reincarnation of Section 6.3. Based on the property that the body of a loop cannot be instantaneous, it ensures that no control arc of the flowgraph is traversed by control more than once in an instant.

Our translation scheme may replicate code even when doing so is semantically unnecessary. Static analysis techniques like those used in the circuit-based INRIA compiler and the Esterel V7 compiler of Esterel Studio can reduce the degree of replication. We preferred the current solution because it is simple to define and implement, it allows the definition of a simple link between the circuit translation and the compilation schemes defined in the following chapters, and it greatly simplifies the control and data dependency system and hence the optimization, encoding, and scheduling into sequential code.

8.2.2 Translation Rules

Pause

The translation of **pause**, Figure 8.8, maps directly to the semantics of the statement. When started, the status of the statement is 1 (selected) and completion code 1 is generated. When selected and resumed, the status of the statement is 0 and completion code 0 is generated. Since **pause** does not read or emit signals, no signal link sets are defined.

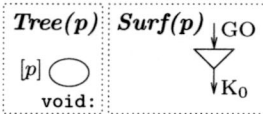

Figure 8.9: The translation of $p=$"nothing" and $p=$"emit S"

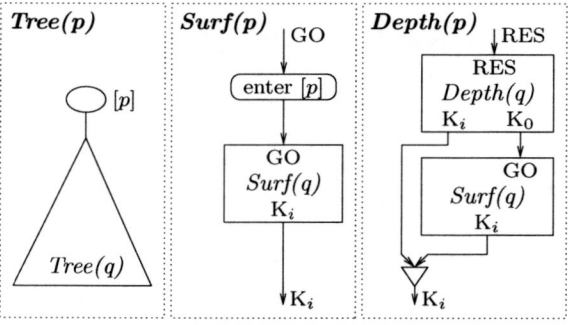

Figure 8.10: The translation of $p=$"loop q end"

Nothing

The statement **nothing** is instantaneous, so it only generates a surface graph, given in Figure 8.9. Since **nothing** does not read or emit signals, no signal link sets are defined.

Loop

The translation of the loop primitive follows the translation and reincarnation technique used in the circuit translation (cf. Section 6.3, Figures 6.14 and 6.15). The associated GRC is presented in Figure 8.10. The signal link sets are:

$$\begin{aligned} SurfIn(S)(p) &= SurfIn(S)(q) \\ SurfOut(S)(p) &= SurfOut(S)(q) \\ DepthIn(S)(p) &= DepthIn(S)(q) \cup SurfIn(S)(q) \\ DepthOut(S)(p) &= DepthOut(S)(q) \cup SurfOut(S)(q) \end{aligned}$$

Sequence

The GRC translation of the sequence (Figure 8.11) follows the same reincarnation pattern. The pattern in the figure is for the binary sequence; using it

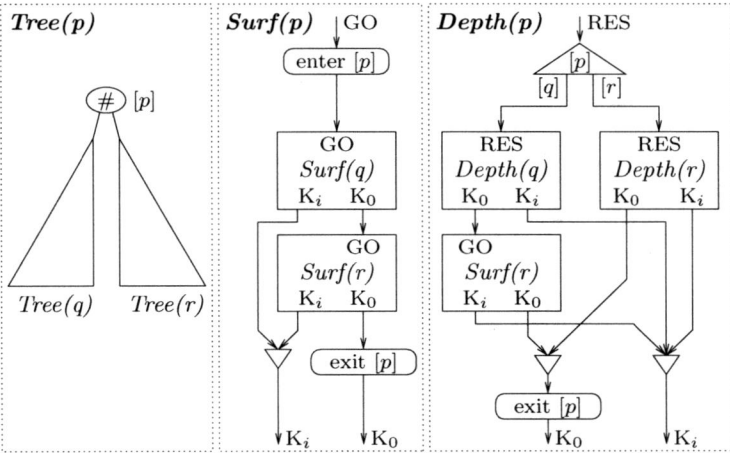

Figure 8.11: The translation of $p=$ "q ; r." The completion code index i ranges over the potential completion codes of p, except 0 (cf. Section 4.6).

to translate n-ary sequences is inefficient—code duplication may be $O(n^2)$. Instead, this pattern is easily extended so the surface of each branch is duplicated at most once (compilers implement this optimized version). The signal link sets are obtained by uniting the signal link sets of the composed sub-graphs:

$$\begin{aligned} \mathit{SurfIn}(S)(p) &= \mathit{SurfIn}(S)(q) \cup \mathit{SurfIn}(S)(r) \\ \mathit{SurfOut}(S)(p) &= \mathit{SurfOut}(S)(q) \cup \mathit{SurfOut}(S)(r) \\ \mathit{DepthIn}(S)(p) &= \mathit{DepthIn}(S)(q) \cup \mathit{DepthIn}(S)(r) \cup \mathit{SurfIn}(S)(r) \\ \mathit{DepthOut}(S)(p) &= \mathit{DepthOut}(S)(q) \cup \mathit{DepthOut}(S)(r) \cup \mathit{SurfOut}(S)(r) \end{aligned}$$

Parallel

Figure 8.12 is the code generated for $p=q \parallel r$. The **Sync** nodes gather the completion codes from the branches. In the depth, they also gather the Dead() wires that are activated if the corresponding branches have terminated in previous instants. In the depth graph, selection tests verify if the branch code must receive control. The signal link sets are obtained by uniting the signal link sets of the composed sub-graphs:

$$\begin{aligned} \mathit{SurfIn}(S)(p) &= \mathit{SurfIn}(S)(q) \cup \mathit{SurfIn}(S)(r) \\ \mathit{SurfOut}(S)(p) &= \mathit{SurfOut}(S)(q) \cup \mathit{SurfOut}(S)(r) \\ \mathit{DepthIn}(S)(p) &= \mathit{DepthIn}(S)(q) \cup \mathit{DepthIn}(S)(r) \\ \mathit{DepthOut}(S)(p) &= \mathit{DepthOut}(S)(q) \cup \mathit{DepthOut}(S)(r) \end{aligned}$$

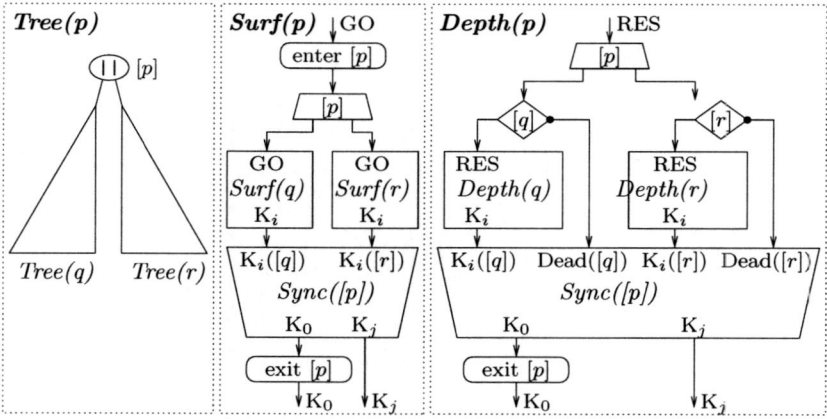

Figure 8.12: The translation of $p=$"$q \parallel r$." Index i ranges over the potential completion codes of the branches. Index j ranges over the potential completion codes of the parallel, except 0.

Signal declaration and emission

The emit statement is instantaneous, so it only generates a surface graph like the nothing statement (Figure 8.9). Only the $SurfOut(S)(p) = \{J.CONT\}$ signal link set is not empty, where $J.CONT$ is the $CONT$ output port of the Join node generated for p (the K_0 output port of the pattern).

The generation of the GRC for the signal declaration statement starts by encapsulating the GRC of its body in a shell by adding the extra selection node and corresponding state update calls. Figure 8.13 shows this.

The second translation phase performs the signal link. In the surface

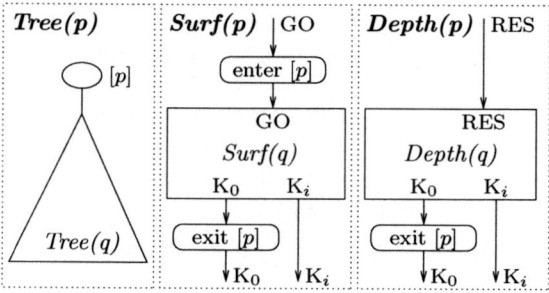

Figure 8.13: The translation of $p=$"signal S in q end." Index i ranges over the potential completion codes of p except 0.

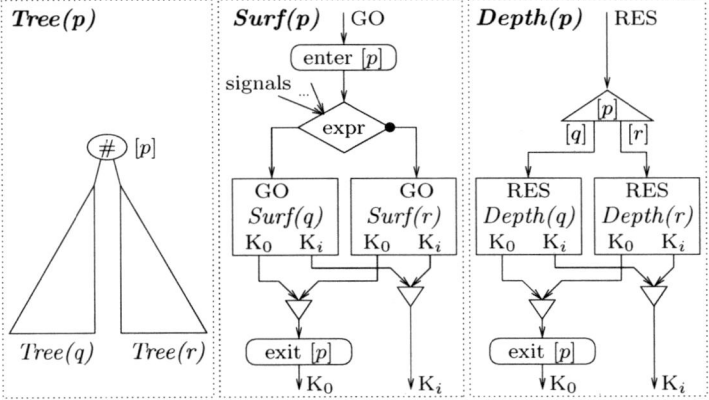

Figure 8.14: The translation of $p=$"present S then q else r end" and $p=$"if v then q else r end." Index i ranges over the potential completion codes of p except 0.

graph, it adds a control arc between every output port of $SurfOut(S)(q)$ and every input port of $SurfIn(S)(q)$. In the depth graph, it adds a control arc between every output port of $DepthOut(S)(p)$ and every input port of $DepthIn(S)(p)$. All the signal link sets corresponding to signals $U \neq S$ are transmitted unchanged from q to p. The signal link sets corresponding to S are all set to $\emptyset$ in p.

Test statements

Figure 8.14 shows the GRC generated for signal and data test statements. The difference between the two appears in the test expression and the signal input ports on the **Test** node.

For a signal test, the **Test** node has an input port for the test signal S. For a test on the sequential variable v, the expression is the test action test_action(v), as defined in Section 6.3.2. The four signal link sets for a variable test statement are the union of the corresponding branch sets, for all signals U:

$$\begin{aligned}
SurfIn(U)(p) &= SurfIn(U)(q) \cup SurfIn(U)(r) \\
SurfOut(U)(p) &= SurfOut(U)(q) \cup SurfOut(U)(r) \\
DepthIn(U)(p) &= DepthIn(U)(q) \cup DepthIn(U)(r) \\
DepthOut(U)(p) &= DepthOut(U)(q) \cup DepthOut(U)(r)
\end{aligned}$$

In the case of the signal test, only the definition of $SurfIn(S)(p)$ changes:

$$SurfIn(S)(p) = SurfIn(S)(q) \cup SurfIn(S)(r) \cup \{T.\text{IN}(S)\}$$

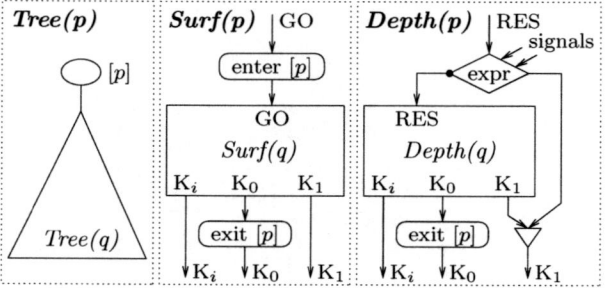

Figure 8.15: The translation of $p=$"suspend q when $expr$." Index i ranges over the potential completion codes of p except 0 and 1.

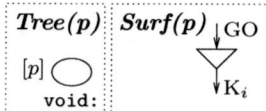

Figure 8.16: The translation of $p=$"exit $T(i)$"

where $T.\text{IN}(S)$ is the IN(S) input port of the test node T.

Suspend

Figure 8.15 shows the GRC for the **suspend** statement. All the signal link sets, with the exception of $DepthIn(S)(q)$, are passed unchanged from q to p. The exception set is

$$DepthIn(S)(p) = DepthIn(S)(q) \cup \{T.\text{IN}(S)\},$$

where $T.\text{IN}(S)$ is the IN(S) input port of the generated test node T.

Trap exit and handling

The **exit** statement is instantaneous. It only generates a surface graph (Figure 8.16). The signal link sets are empty.

Figure 8.17 shows the GRC for the trap statement. Like the circuit translation pattern, the GRC pattern converts completion code 2 to normal termination (code 0), decrements the remaining trap codes, and resets the state of the statement when the body completes with code 2. The signal link sets are transmitted unchanged from q to p.

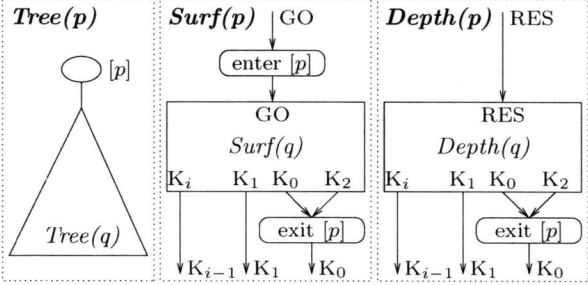

Figure 8.17: The translation of $p=$ "trap T in q end." Index i ranges over the potential completion codes of q, except 0, 1 and 2.

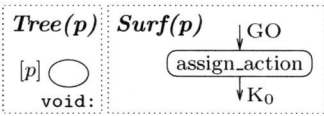

Figure 8.18: The translation of a variable assignment statement p

Data assignment

Variable assignment statements are instantaneous. Therefore, we only need the surface graph: Figure 8.18. According to the data encoding defined in Section 6.3.2, the assign action is svar_emit(s) if $p=$ "$s <= f(\ldots)$" and var_assign(v) if $p=$ "$v := f(\ldots)$." The signal link sets are empty.

Variable declaration statements

Figure 8.19 shows the GRC for variable declaration statements. The initializing action is performed just before starting the body statement p and the reset action is performed just before resuming it.

If the translated statement is $p=$ "shared s in q end," then the initialization action is shared_init(s) and the reset action is shared_reset(s). When $p=$ "var v in q end," then the initialization action is var_init(v) and the reset action does nothing.

When the shared variable declaration "shared s in q end" is translated into GRC, we also need to generate the data dependency arcs linking the actions that write s with the actions that read s. Linking is done separately for the surface and depth graphs. In each case, one data dependency arc is generated from every call of svar_emit(s) to every action that reads s.

All the signal link sets are passed from q to p unchanged.

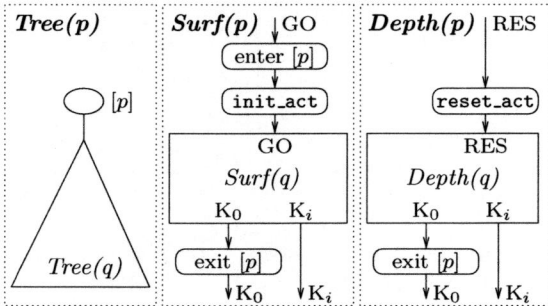

Figure 8.19: The translation of a variable declaration statement p of body statement q. Index i ranges over the potential completion codes of p except 0.

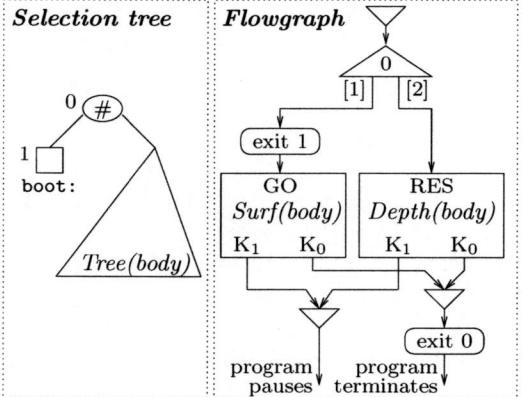

Figure 8.20: The global translation context

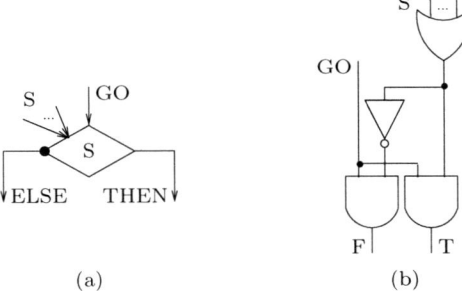

Figure 8.21: Circuit counterpart for the **Test** node

8.2.3 The Global Context

After generating surface and depth code, which define the behavior of the program at different execution instants, we have to link them into a single flowgraph representing the global behavior of the program. To do so, the selection tree and the surface and depth code are placed into the context of Figure 8.20. The boot selection node is selected to mark the program start instant.

8.3 Formal Simulation Semantics and Translation Correctness

While our intermediate format has a control flow graph as its central component, classical control flow semantics cannot handle the full semantics of Esterel(see our arguments in Section 3.1).

To correctly represent the correct semantics of the programs, the GRC flowgraph is given constructive operational semantics by extending the previously defined semantics of the circuits with data. The computation of an instant is an evaluation process where the control arcs (like the circuit wires) change their status bits from $\bot$ (undefined) to either 0 or 1. The evaluation of certain nodes require the execution of data actions. Data arcs impose supplementary causal constraints on the execution of data actions. The program is incorrect (not constructive) if we cannot set every arc to a value of 0 or 1. This evaluation policy is general enough to support the evaluation of mixed circuit-flowgraph specifications.

To define the function of the GRC nodes, we give circuit counterparts that perform the same computation under constructive semantics. This not only simplifies the definition, which can be complex, but also emphasizes the strong link between the two representations. Figures 8.21 and 8.22 give the circuit counterparts of the complex GRC nodes.

The circuit associated to **Sync** is the circuit parallel synchronizer in Figure 6.20. The **Tick** do not generate gates.

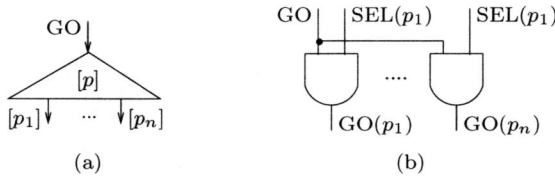

Figure 8.22: Circuit counterpart for the **Switch** node

State encoding is the only difference between the circuits obtained by circuit expansion of the GRC flowgraph and those generated by directly following the rules of Section 6.3. In GRC-generated circuits, all current selection status bits are circuit inputs since the selection tree is not present. Their computation is performed using the Boolean selection status bits (variables) and the enter/exit primitives. At the beginning of each instant, the selection circuit inputs are set according to the associated selection status bits. Note that this implies that the state of each statement is decoded at each instant prior to program execution. While for performance reasons this can be avoided at code generation time for most programs, Section 9.3 shall present a (cyclic) Esterel example whose simulation requires it.

Translation correctness

The circuit expansion of GRC and the direct circuit translation of a given program result in circuits with data that are identical up to the state representation. Then, the simulation of an instant for given state and input event produces the same output signals in both circuits and leaves user data with the same values.

Therefore, proving the full equivalence between the GRC-based circuit expansion and the direct expansion consists in proving that the following two properties are true in the end of each execution instant where the simulation set a value of 0 or 1 to every circuit wire: the output selection flags produced by the evaluation of the GRC-generated circuit are consistent and the flags corresponding to the "pause" statements are equal to the input values of the corresponding registers.

If we assume that the circuit translation of Chapter 6 is correct with respect to the direct semantics, the equivalence between the GRC-based circuit translation and the direct circuit translation proves the correctness of the Esterel translation to GRC.

8.4 Format Optimizations

As usual with syntax-directed translation, GRC produced by structural translation from Esterel programs is often far from optimal because subprograms are translated in a way that allows them to operate in any context, not just in their specific context. It is therefore important to optimize the intermediate code. Support for optimization was a main motivation behind the design of GRC.

The problem is not specific to our technique; structural translation to hardware also generates redundant circuits. The use of optimizations is imperative in the generation of efficient circuit code. State-of-the-art tools exist (e.g., SIS [64]) that implement well-studied combinational and sequential optimization techniques like those described by Hachtel and Somenzi [37]. Sentovich, Toma, and Berry [63] and Touati and Berry [70] study the application of such optimizations to circuits generated from Esterel and show impressive results: circuits can shrink to as little as 10% of their original size. See also the Ph.D. theses of Fornari and Toma [69, 32]. This can improve the size and speed of the generated (software) circuit code to the point that it competes with control flow-based compilers. The computational cost of good circuit optimizations is an issue here. These techniques are often global, and may require computing the reachable state space (usually using a BDD), which can be prohibitively costly on large programs. More recently, the Esterel V7 circuit compiler supports modular compilation of Esterel programs to circuits with independent local optimizations for each module, which is less costly but not as effective.

The first control-flow code generators (Synopsys, Saxo-RT) present a different picture: unlike the simple, well-studied mathematical model of digital circuits, their intermediate formats involve complex computation graphs and custom state encodings tailored to the form of the generated code. The intermediate representations include partial C-language mappings, and the compilers usually include fast, software-specific optimizations. The algorithms are similar to those used in the compilation of traditional imperative languages: dead code removal and control and data dependency simplification.

The GRC-based compiler advanced the state of the art by adapting fast existing optimization techniques while defining new, Esterel-specific optimizations based on a deeper understanding of the program semantics. Adapted techniques include dead code removal and sweep-like constant propagation. New Esterel-specific algorithms simplify the state representation. All the GRC optimizations are able to handle cyclic specifications and respect the constructive semantics of the format.

All the algorithms adapted or defined for use with GRC are fast. Their complexity is low-exponent polynomial in the size of the GRC graph. In practice, the full optimization cycle takes less than one minute on our largest example. Some of the techniques can be refined to use "expensive" information, such as the reachable state space structure, but this direction has not been explored. The optimizer is an independent, optional compiler module that is fast enough to use always.

Given that GRC can be seen as a step towards circuit translation, all the optimization techniques defined in this section are general enough to be adapted to optimizing Esterel-generated circuits. While this is obviously true for techniques borrowed from the circuit world, the novel state representation simplifications also give good results, as shown by experiments presented elsewhere [55].

We move now to the actual description of the algorithms. They are grouped in two classes, depending on whether they operate mainly on the selection tree or the flowgraph.

Selection tree analysis determines a number of static activation properties, such as the fact that a given statement never terminates. Once determined, these properties are represented on the selection tree and later used as basis for flowgraph simplifications and, at code generation time, to improve the C language encoding of the state representation.

Flowgraph simplifications are the actual optimizations that reduce the number of nodes, calls, and dependencies.

The two classes of algorithms are not independent since modifications of one structure need to be reflected in the other. Not only must the coherence between between the selection tree and flowgraph be preserved between optimization steps, but results of one optimization class may lead to improvements in the other. The optimization script alternates between the two types of algorithms a fixed number of times.

8.4.1 State Representation Analysis

The selection tree of a GRC specification is an abstraction of the syntax tree of the Esterel source; the hierarchy is enriched with tags representing static selection node properties. Part of the tags, describing the syntactic parallel/exclusive structuring of the state representation, are set in the Esterel to GRC translation—`pause:`, `ref:`, `exclusive:`, and `parallel:`. The remaining two (`nonterm:` and `void:`), which also represent properties with great impact on optimization and code generation, are computed through static analysis of the GRC. The analysis can also lead to the modification of tags generated in the translation process. Below, we explain the meaning of the new tags and how they are computed.

Statements that do not hold state

While translating Esterel programs, it is important to identify which statements do not retain control between instants. Such statements do not require a state representation and are never resumed. Obviously, all the constructions that do not implicitly contain a `pause`, such as "`emit S`" and "`present S then emit O1 end`," fall in this class. With the use of traps, however, instantaneous statements containing `pause` can be constructed. For example,

```
trap T in
  pause
```

```
    ||
       exit T
    end
```

Here, when the first parallel branch pauses, the second branch causes the instantaneous preemption of the entire construct. While we do not expect anyone to write this particular example, composing library modules often produces something similar.

Current analysis techniques are able to identify some of these instantaneous statements. The information is preserved by tagging the associate selection nodes with `void:`.

Nonterminating statements

The following example illustrates the second property that proves important in practice: the non-termination of parallel branches.

```
    [
       sustain S ;
    ||
       await I ; exit T
    ]
```

Here, neither branch terminates (i.e., completes execution with completion code 0). The first always pauses. The execution of the second branch can only finish by raising trap T, in which case the entire parallel statement is preempted. Consequently, the selection status bits for the two branches and that for the parallel statement are always equal. Instead of three bits, we only need one: the one belonging to "await." Furthermore, the state test/update protocol can be simplified, since no selection test is needed in the depth for the parallel branches and the *enter/exit* primitives that access redundant flags are also unnecessary.

This kind of redundancy arises naturally from the Esterel programming style that encourages hierarchical composition of preemption and non-terminating behaviors. Table 8.1 gives some statistics on redundancy in typical programs. The table lists the percentage of parallel branches that were found nonterminating according to an algorithm presented below. The figures could be higher since these algorithms are only approximate (they do not take signaling into account). We mark the selection nodes corresponding to parallel branches that cannot terminate with `nonterm:`.

Tag computation

The best source of information about the behavior of the selection tree is the reachable state space of the Esterel program, but its computation can be intractable even for medium-size examples. We need cheaper ways to compute the tags.

Example	Percentage of nonterminating branches
1-tcint	97%
2-wristwatch	97%
6-fuel	98%
7-cabine	87%
8-global	70%

Table 8.1: Parallel branch status redundancy statistics

Establishing `nonterm:` flags requires us to identify parallel branches that never terminate. The current technique is based on a simple analysis of GRC **Sync** nodes. Let q be a branch of the parallel statement p. Through translation, p produces a number of **Sync** nodes that have as inputs completion code arcs output by q. If none of these **Sync** nodes receives an arc corresponding to completion code K_0 from q, we know q cannot terminate and mark its selection node with "`nonterm:`." This analysis is a form of potential propagation, which gives a static approximation of the dynamic *Can* computation.

We also mark the root selection node with `nonterm:` if we determine that the program cannot terminate (i.e., if there are no more calls to "`exit 0`").

Note that the tagging process should be repeated after each flowgraph optimization step because removing completion code arcs may reveal new non-terminating parallel branches.

The computation of `void:` flags relies on several criteria. We mark the following with `void:`.

- Nodes that are not `pause:` and that only have `void:` children or no child at all. These are the composed statements with instantaneous bodies/branches.

- Parallel nodes with a child tagged with both `void:` and `nonterm:`. In this case falls "`pause || exit T`," since the second branch is both instantaneous, and cannot terminate.

- Nodes for which no **enter** primitive exists within the control-flow graph. They correspond to dead code—statements that contain "pause" statements but are never started, e.g., "`exit T ; pause`."

Existing tags can be updated to reflect finer understanding of the selection tree properties. In particular, the tags `exclusive:` and `parallel:` are not interesting and deleted from nodes that have at most one non-void child. The exact transformations are the following.

- When the `void:` tag is set on a selection node, the existing `pause:`, `ref:`, `exclusive:`, or `parallel:` tags are deleted. Nodes tagged with

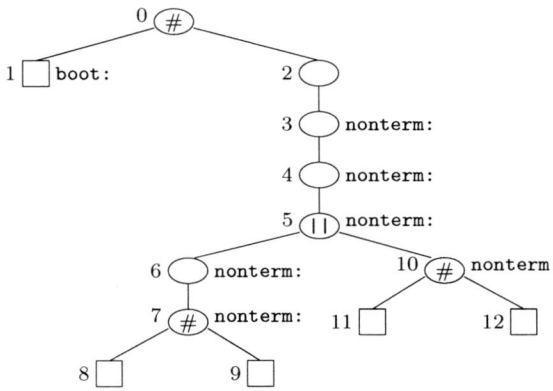

Figure 8.23: Tagged tree for MainExample

void: do not have a state representation so the other tags are meaningless.

- When a node tagged with **parallel:** or **exclusive:** has exactly one non-void child, the existing tag is replaced with **ref:**. Again, this reflects the fact that **void:** selection nodes do not take part in the actual encoding.

The computational complexity of the algorithms in this section is linear in the size of the GRC specification. Figure 8.23 presents the selection tree of MainExample after tagging.

8.4.2 Flowgraph Optimizations

Simplifications of the GRC (removal of arcs and nodes) are performed at the level of the control/data flowgraph. "Simplification" describes the approach better than "optimization" since no complex re-encoding is performed to reduce the redundancy of the specification. The algorithms described below only remove nodes, calls, and signal/data dependencies, but otherwise do not change the topology of the graph.

The graph simplification algorithms are divided in two processing phases, each exploiting a different source of information. First, the flowgraph is trimmed taking into account properties stored in the selection tree as described above. Then, a complex propagation mechanism finds and removes internal flowgraph redundancies without analyzing or changing the selection tree.

Selection-based simplifications

The following four simplifications are performed based on the selection tags.

- Calls to selection primitives on `void`: selection nodes are deleted. In the case of `test` primitives, we also delete the calling **Test** node and replace it with a **Join** node linking the input *GO* port with the *ELSE* port of the test.

- The `test` primitive calls on `nonterm`: selection nodes are deleted. The calling **Test** nodes are replaced with **Join** nodes linking the ports *GO* and *THEN*.

- The **Switch** nodes are simplified by erasing the outputs corresponding to `void`: branches (such nodes cannot receive control, so their code can be erased). If only one output remains, the `switch` call and **Switch** node are deleted.

- The `sync` calls on parallel nodes that have at least one `nonterm`: child are deleted. Such calls were used to determine whether or not unfinished branches still exist, in order to determine the status of the parallel statement itself. As the status of the parallel is known to be equal to any of the `nonterm`: branches, we no more need these calls.

In addition to nodes deleted using the `nonterm`: and `void`: tags, we can also exploit the hierarchical structure of the selection tree and delete all the `enter` and `exit` calls situated on a parallel branch whose output leads to no $K_i(br)$ wires with $i \leq 1$. State operations are useless in this case since in the end they are overridden by some higher-level trap. Note that this simplification can only be applied if the corresponding parallel synchronizer has not been optimized out.

Selection-independent simplifications

In the second phase we perform three simplifications that do not use information from the selection tree.

The first is the so-called constructive sweep that corresponds to a classical dead code sweep. It consists of determining, through static propagation of *false* values, which arcs are always set to *false*. These arcs can be deleted and so can be some of the adjacent nodes. More precisely, we delete **Call**, **Test**, **Switch**, and **Join** nodes whose *GO* input port no longer admits incoming arcs, as well as **Sync** nodes left without incoming arcs (dead code); and **Test** nodes whose expression has been uniformly set to *true* or *false* when input signals are no longer emitted. For instance, in the following fragment

```
signal S in
  pause ; emit S ;
  pause ; present S then emit O1 else emit O2 end
end
```

the code corresponding to the "present" statement is simplified to "emit O2."

Useless code simplification consists of deleting code that does not drive actions on data (state changes, output signal emissions, or user data update). For instance, **Join** and **Sync** nodes whose outputs are not connected to other nodes can be deleted. In the code generated for the following program, the **Sync** nodes can be deleted because they perform an unnecessary function.

```
module simple:
output O1, O2;
[ sustain O1 || sustain O2 ]
end
```

Dependency simplification consists of removing signal and data dependencies between nodes belonging to exclusive control threads. This uses exclusivity relations on the outputs of a **Test**, **Switch**, or **Sync**. The simplest case concerns the **Switch**. In every circumstance, we can remove signal and data dependencies between statements separated by **pause** statements. In the following example, the automatic translation scheme will link with a signal arc the emission of S with its test.

```
pause; emit S ;
pause; present S then emit O1 end
```

This is a false dependency that can be removed. Note that the way we encoded dependencies, defined in Section 8.1.2, greatly influences the ability to perform this optimization. In our case, the false dependency can always be removed, as its representation only involves the emitting and the receiver port. In the Esterel V7 compiler, this is not the case. There, hub nodes gather all emissions of a signal before distributing it to all reception ports (Figure 8.5). Consequently, the ability to remove a false dependency depends on the context. For instance, putting the previous statement in parallel with the next one makes dependency removal impossible.

```
pause; present S then emit O2 end ;
pause; emit S ;
```

The situation is more complex for tests. For instance, if I is an input signal, we can remove the dependency between the emission of S and its test in the following example.

```
present I then
  emit S
else
  present S then emit O1 else emit O2 end
end
```

Removing the dependency is incorrect in the general case, as the following fragment shows.

```
signal S,T,U,V in
  present V then
    present U then
      present S then emit T ; emit U ; emit V end
    else
      present T then emit S ; emit U ; emit V end
    end
  end
end
```

The example is non-constructive; there is no way to decide the status of the signals S, T, and U without speculation. However, removing the dependencies on S and T makes the GRC constructive (even acyclic, if a pass of constructive sweep is performed).

The previous non-constructive example is logically correct in the sense of Berry [7], meaning that in classical logic its system of logical relations has a single solution: all signals absent. Moreover, all non-constructive examples whose GRC becomes constructive through dependency simplification are logically correct.* Based on this argument, the Saxo-RT compiler accepts programs such as our last example. Our choice is to reject such programs, because accepting them changes the causality and transforms program correctness into an algorithmic issue depending on compiler internals.

Consequently, we restrict the use of the dependency simplification to cases where it is not dangerous. Dependency simplification is correct

1. between nodes situated on different branches of a data test when at least one of the considered nodes is not on the same strongly connected component (SCC)† of the GRC flowgraph as the test node;

2. between branches of a signal test if none of the signal wires are part of the same cycle as the test node itself, or if the node is not part of any cycle;

3. between branches of a state test (this case is not interesting since the current code generation scheme never puts code on the *false* branches of such tests); or

4. between output branches of a **Sync** node if it is not part of a cycle.

We did not explore other cases where dependency simplifications can be performed. Other simple cases exist where we could use them, such as in the following fragment where the dependency on T is useless when S is an input signal.

---

*Proof sketch: Removing the dependency amounts to using the law of the excluded middle to simplify the logical expression of the test node outputs. Applying the excluded middle does not change the set of Boolean fixpoints of the program/circuit.

†A Strongly Connected Component (SCC) of a directed graph is a maximal set of nodes such that there is a path connecting in any order any two of its nodes.

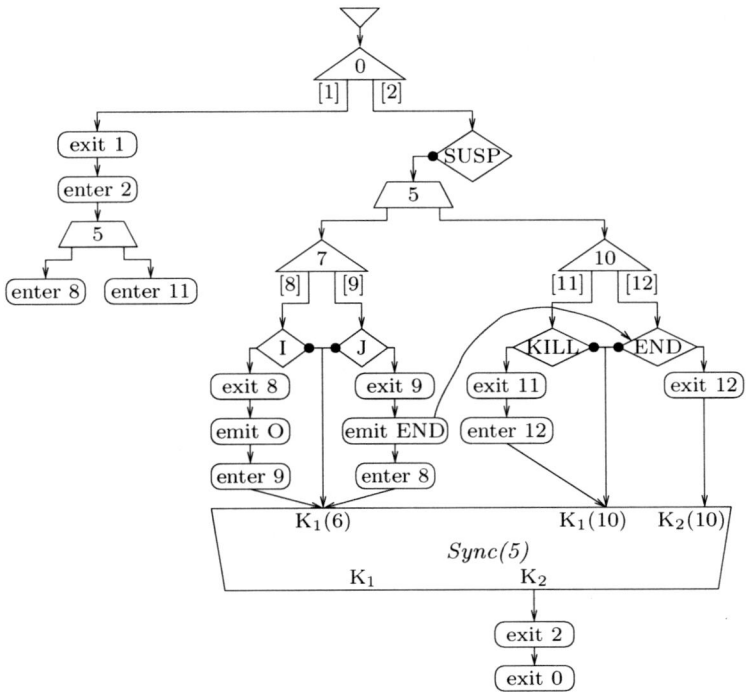

Figure 8.24: Simplified graph for `MainExample`

```
[
  present S then emit T end
||
  present S else present T then emit O1 else emit O2 end
]
```

Figure 8.24 shows the result of simplifying the control-flow graph of `MainExample`. The selection tests corresponding to `nonterm:` branches have disappeared, as have the corresponding *enter* and *exit* primitive calls. The interface of the **Sync** node was simplified and the *sync* call has disappeared.

9
Code Generation from GRC

In this chapter, we present the technique used in the INRIA compiler to generate efficient sequential (C) code from GRC specifications. In this compiler, dubbed grc2c, code generation is based on statically scheduling the GRC flowgraph nodes into well-structured sequential code. In this sense, grc2c follows the approach of the Synopsys and Saxo-RT compilers described in Chapter 7, but pushes the efficient encoding of the state and reactive operations further.

Static scheduling is possible only when the GRC flowgraph is acyclic. The notion of an acyclic intermediate representation is therefore central in this code generation scheme, just as it is in the other circuit and control-flow code generators. This is why we dedicate the first part of this chapter to better understand what "acyclic" means at both circuit and GRC levels. Based on this analysis, we are able to view acyclicity independently from particular intermediate representations (GRC, netlist, etc.), making it a well-defined correctness criterion.

The second part of the chapter defines the code generation technique, which is based on static scheduling.

The last part of the chapter defines an extension of the static scheduling approach that allows code generation for all Esterel programs that are syntactically correct. This generates code that performs dynamic scheduling of each strongly connected component (SCC) in the flowgraph. The code associated to an SCC is embedded in the (now) globally acyclic graph and triggered by the latter during reactions where control traverses the SCC. Static scheduling can now be applied to the global GRC flowgraph. For simplicity, the semantic evaluation is performed at the circuit level, as explained in Section 8.3.

9.1 Defining "Acyclic"

The notion of a cyclic specification is central to most Esterel compilation techniques, including the control flow- and circuit-based techniques. Here, we explain the importance of the notion, emphasize its problems, and then propose a partial solution.

Section 2.3 discussed an intuitive source-level definition of "cyclic." There, signal and data dependencies may chain into static causality cycles that complicate the execution by requiring dynamic scheduling. Cyclic specifications may have semantic problems since some are non-constructive.

Formally, cycles are defined at the level of intermediate graph-based formats—digital netlists or the intermediate format of one of the control-flow code generators. For instance, we say that a GRC specification is acyclic if its flowgraph is acyclic.

Acyclic representations have specific qualities. First, acyclicity at any of the above-mentioned levels enforces correctness with respect to constructive causality. Moreover, acyclicity at the circuit level is equivalent to constructiveness when the lazy ternary logic operators of Figure 3.3 are replaced with their strict counterparts. As most real-life programs generate acyclic intermediate representations, it is easy to declare them correct without relying on expensive reachable state space exploration algorithms [69]. Also, acyclic specifications can be statically scheduled into fast code. Chapter 7 described how some control-flow code generators do this. The circuit code generators also produce better (faster) code for acyclic circuits [32].

Unfortunately, the definition of acyclicity is a problem because it is not a property of program syntax, and the flowgraph that must be acyclic varies according to details of the translation scheme. It is then very difficult to compare the classes of programs accepted by different compilation techniques. Each existing Esterel compiler accepts a slightly different class of programs.

The Esterel example of Figure 9.1(a), for instance, has a cyclic GRC representation (Figure 9.1(b)), but an acyclic circuit (Figure 9.1(c)). This occurs because of the finer decomposition of the GRC-level parallel synchronizer into gates, despite similar translation patterns.

Further optimizations may also interfere with acyclicity properties, as the following example shows.

```
pause ; present S then emit T end ;
pause ; present T then emit S end
```

Both the circuit and GRC generation schemes produce cyclic representations for this program. However, the optimization techniques we defined earlier can break the cycles by observing that the two tests are never executed in the same instant. The INRIA, Synopsys, and Saxo-RT control-flow code generators implement this optimization, but not the INRIA circuit code generator.

Our goal is to define acyclicity in a compiler-independent way. Among the existing intermediate formats, digital circuits have a privileged position

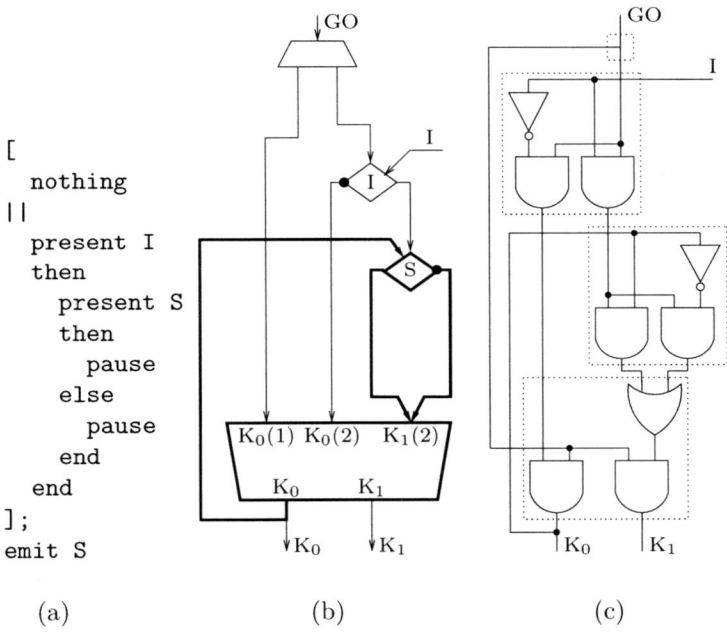

Figure 9.1: Esterel program (a) for which the GRC-level cycle (b, emphasized) is resolved at the circuit level (c). The dotted boxes mark the correspondence between GRC nodes and netlist gates. State encoding/decoding components have been omitted because they never determine cycles.

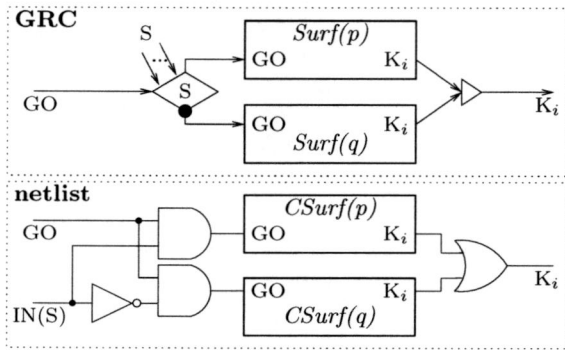

Figure 9.2: Simplified surface translation patterns for **present**

in this respect. Having the finest-grain representation, they offer a notion of acyclicity that is only approximated by the other formats (GRC included). Intuitively, a program generating an acyclic GRC representation should always produce an acyclic circuit. When this does not happen in practice, the cause is either a significant variation of the structural translation (reincarnation) pattern or a more aggressive optimization of the GRC code that is not applied on the circuit.

In the sequel, we shall compare the circuit and GRC translations from this point of view. We find the acyclicity of the GRC flowgraph implies an acyclic circuit. The strong topological similarity of the two schemes also allows us to identify a simple refinement of the basic GRC that unifies the two versions of acyclicity. The equivalence is preserved if the (refined) GRC and the circuit are both optimized using techniques described in the previous chapter.

This equivalence result is weak in its formulation, since it concerns particular translations. Nevertheless it can be useful in practice because it offers a roadmap for developing "compatible" circuit and control-flow code generators that accept the same classes of programs.

GRC format refinement

In circuits and GRC specifications derived from Esterel programs, we can identify a class of components whose only role is to encode the state of the program in the next instant. In GRC, these components are the *enter* and *exit* calls; the RIN wires and their driving gates perform this function in circuits. The state encoding components are never involved in cycles. Therefore, we can perform our analysis using simplified translation patterns, such as the one in Figure 9.2. Figure 9.1 shows an example of translation using the simplified patterns.

In circuits, we shall also overlook gates that compute the SEL wires starting from the register outputs. These gates, organized in a tree with the register outputs as leaves, also cannot be part of a cycle.

After the simplification of unneeded state encoding components, a simple correspondence exists between GRC nodes and arcs and circuit gates and wires. In fact, the Esterel translation into circuits can now be factored into the Esterel translation to GRC followed by the expansion of the nodes into gates, as defined in Section 8.3. By expanding the nodes of the simplified GRC code into gates, we obtain a circuit that is identical to the circuit translation except for wires and gates that correspond to Esterel-level signal communication. The difference comes from the fact that the GRC translation uses no "hub" OR gates to group signal wires before distributing them. Instead, individual wires (the signal control arcs) connect every emission place with every test. Note that the result is the same with respect to causal emission/reception dependencies, at least until optimizations are applied.

Figure 9.1 is an example of the mapping. There, gates associated with specific nodes are grouped in dashed boxes.

The simplified GRC is an abstracted version of the circuit, so an Esterel program that produces an acyclic GRC specification will produce an acyclic circuit. However the converse is not true. Below we discuss when an Esterel program can become an acyclic circuit from a cyclic GRC specification.

Cycles disappear from the dependency graph only when GRC parallel synchronizer **Sync** nodes are expanded into gates; expanding other types of nodes does not affect the cyclicity of the circuit. The analysis is performed case by case, according to node type.

- **Tick** nodes generate no circuit components.

- **Call** nodes are translated into buffer gates that preserve the GRC-level causal dependencies between input and output.

- Each **Test** node is translated into two gates, but these preserve the GRC-level causal dependencies of one of the test node output ports.

- **Switch** nodes, like the test nodes, preserve through expansion all the GRC-level causal dependencies between inputs and outputs.

Thus, only expanding **Sync** nodes can remove causal dependencies and eliminate cycles. Unfortunately, the circuit expansion of a synchronizer is complex (Figure 6.20) and thus best avoided when generating sequential code.

Here, we propose a technique that allows us to split the synchronizers into a minimal number of components, which can thereafter be easily encoded into software sequential code. This technique is useful when the circuit is acyclic but the GRC is not.

Our minimal refinement follows from the internal structure of the parallel synchronizers: the static dependencies between synchronizer ports that disappear after circuit expansion are $K_i(p_j) \to K_n$ with $n < i$. Every other

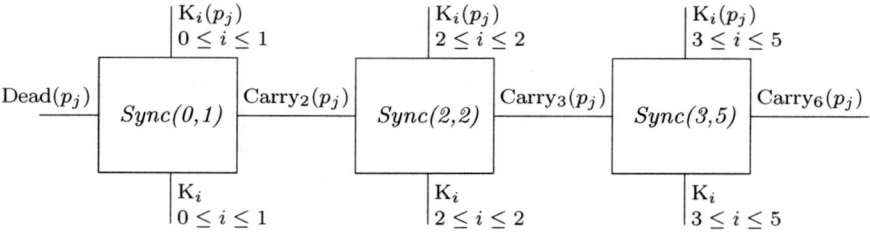

Figure 9.3: Example of synchronizer refinement. The index i ranges over completion codes of the parallel, and p_j ranges over the selection indices of the parallel branches.

dependency is preserved, namely $K_i(p_j) \to K_n$ for $n \geq i$ and $\text{Dead}(p_j) \to K_n$. The indices i and n range here over possible completion codes; p_j ranges over the selection indices of the parallel branches.

When a GRC synchronizer is part of a cycle, static dependencies link synchronizer outputs to synchronizer inputs: $K_n \to K_i(p_j)$ for some i and n. If the circuit expansion breaks the cycle, then $n < i$ for every such dependency. Otherwise, the cycle cannot be broken by circuit expansion. Thus, we are looking for the minimal expansion of a synchronizer under constraints of the form $K_n \to K_i(p_j)$ with $n < i$.

We will split the synchronizers into parts that correspond to sets of consecutive completion codes. Consider a parallel synchronizer **Sync** handling completion codes from 0 to m. For the set of splitting completion codes $0 < k_1 < \ldots < k_r \leq m$ we shall divide **Sync** into the $r+1$ partial synchronizers $\textbf{Sync}(0, k_1-1), \ldots, \textbf{Sync}(k_r, m)$. The partial synchronizer $\textbf{Sync}(a, b)$, $a \leq b$, handles the completion codes from a to b. Its inputs are the completion code wires $K_n(p_j)$, for $a \leq n \leq b$; if $a = 0$, the wires $\text{Dead}(p_j)$; or if $a \neq 0$, the carry wires $\text{Carry}_a(p_j)$. Its outputs are K_n for $a \leq n \leq b$ and $\text{Carry}_{b+1}(p_j)$.

Thus, our problem is to determine a minimal sequence $0 < k_1 < \cdots < k_r \leq m$ such that for each static dependency $K_n \to K_i(p_j)$ ($n < i$) that determines a cycle there exists $n \leq k_t \leq i$. The algorithm that determines this list is given next. It takes as parameter the set of dependencies DEPEND and returns in LIST the list of splitting completion codes. The set DEPEND is filled in with all the pairs (n,i), $n < i$, such that $K_n \to K_i(p_j)$ for some parallel branch index p_j.

```
LIST = empty list
for i = 0 to m do
    if (n, i) ∈ DEPEND for some n then
        append i to LIST
        remove broken dependencies from DEPEND
```

The proof of minimality for this algorithm rests on the fact that split points are only appended to the list where they are necessary. For instance, if only one dependency exists (such as in Figure 9.1), the synchronizer must be split. A more complex example is given in Figure 9.3. There, the synchronizer is split according to the dependency set $\{K_0 \rightarrow K_2(p_1), K_1 \rightarrow K_3(p_1), K_2 \rightarrow K_3(p_0)\}$.

9.2 Code Generation for Acyclic Specifications

We now present the technique used in the INRIA compiler to generate control-flow code from acyclic GRC specifications. It employs state encoding and greedy static scheduling of the flowgraph to generate sequential code with near-automaton speed and size similar to that of circuit code.

An acyclic specification allows the evaluation of GRC nodes to be scheduled statically. Each such schedule satisfies control flow and signal dependencies. In particular, the value of any signal can be read only after all the nodes that can emit it have been evaluated. Thus, no complex "propagation of non-execution" is needed; we can replace the constructive evaluation scheme with simple concurrent control-flow propagation. To generate sequential code, the "threads" defined by the control flow arcs are statically interleaved, using variables to perform the context switches.

The translation is performed in two steps. First, the static properties of the selection tree are used to replace the generic state representation of the GRC specification with a hierarchical bitwise representation on actual C variables. This step also determines a transformation of the flowgraph where the numerous state test and update primitives are replaced by a smaller number of data variable accesses. Every operation on the encoded state representation generally performs several primitive operations.

The second translation step consists of the actual scheduling of the control/data flowgraph into sequential code and C language structures.

9.2.1 State Encoding

The state encoding technique is an improved version of Edwards's [27]. Here, we define it and compare it with the state encoding used in the circuit translation and software circuit code.

We use a hierarchical state representation instead of a flat one such as that used in the circuit translation. In circuits, the memory elements (registers) correspond to "pause" statements. In our representation, the memory elements (C variable bits) correspond to decisions in the control-flow graph. Deciding where memory elements—sequential variable bits—are needed, and how many, is based on the analysis of the selection tree.

In our case, state decoding is embedded in the top-down resumption process. Statements that are suspended or not selected do not receive control and thus do not decode their state because in acyclic specifications, the status

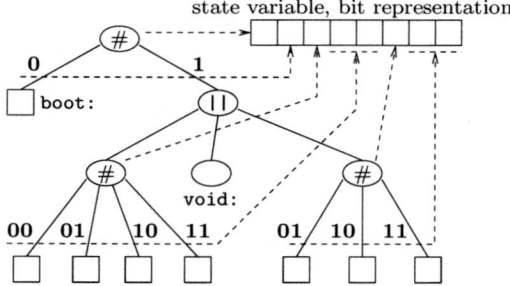

Figure 9.4: Bit allocation for our first small example

of a statement is decoded by the control flow, using a *test* or *switch* primitive, only after the status of each of its parents has been decoded (and found in each case to be *true*).

The re-encoding of the next state works on the same set of state variables as the decoding process. This is possible because all GRC graphs from Esterel are well-formed, so a state update primitive "*enter [p]*" or "*exit [p]*" is only called when no more state decoding primitive can be executed on *Tree(p)*. Thus, the state update primitives *enter* and *exit* can safely make their local updates on the state variables without changing the outcome of subsequent state tests.

This contrasts with the state encoding in the circuit translation. In circuits, the value of every register is recomputed instant, even for those registers corresponding to inactive statements. State decoding works bottom-up starting with the registers and proceeding through the OR gates of the selection circuit and the various selection status wires. There is one further difference in the software code: two full sets of state variables are maintained: one for the current state and one for the next. At the end of each execution instant, the next state is copied onto the current state.

First example

Our first example shows how a selection tree without nonterm: decorations is encoded. The selection tree, pictured in Figure 9.4 along with its encoding, corresponds to the following Esterel fragment.

```
[
  pause; pause; pause; pause
||
  emit S
||
  pause; pause; pause
]
```

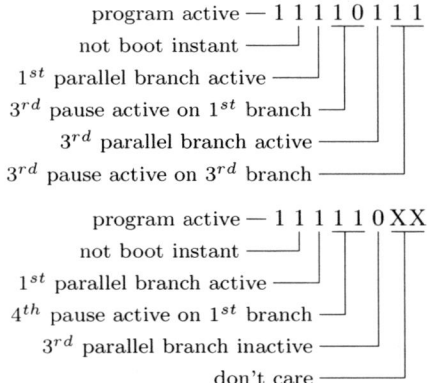

Figure 9.5: State encoding examples for our first small example

Here, the leftmost bit determines whether the program has not terminated, i.e., if the root selection node is selected. If the program is still running, we need another bit to indicate whether we are in the boot instant. If we are not, we need one bit to determine whether the first parallel branch is selected and two bits (encoded logarithmically) to choose the selected pause statement.

The second branch bears the void: tag, but the encoding algorithm ignores it. For the third branch, the simple encoding uses one bit for branch selection and another two to decide which pause is selected. Thus, we need an eight-bit C variable to encode the state of our program. Figure 9.5 gives two actual state encoding examples. Note that not every bit of the state variable is used in every instant.

We could improve the encoding for the third parallel branch. Here, the logarithmic encoding of the selected pause choice leaves the bit configuration 00 unused. This configuration could be used to represent the state where the parallel branch is not selected, so that we only need seven bits to represent the state of our program. However, we have not implemented this optimization. Determining when we can use the incompleteness of a branch representation to represent its selection is not a simple issue, as it may involve a space/speed trade-off, e.g., testing multiple bits that span several C variables is more expensive than testing a single bit.

Second example

The MainExample example on page 147 illustrates how the nonterm: tags are exploited to improve the encoding. Figure 9.6 shows its selection tree and bit assignment.

The leftmost bit indicates whether the program has terminated. The next

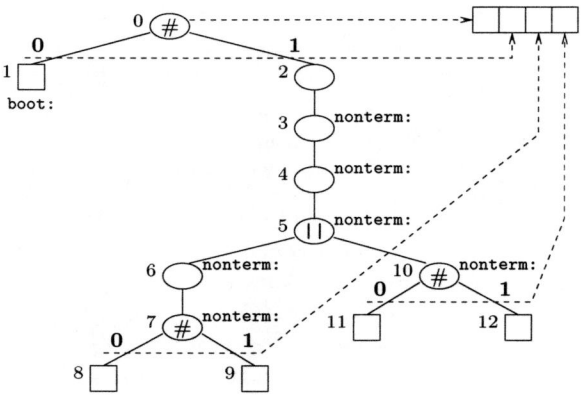

Figure 9.6: Bit allocation for MainExample

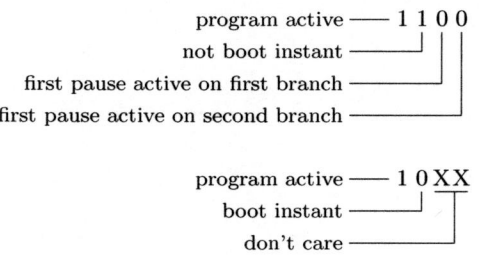

Figure 9.7: State encoding examples for MainExample

one indicates whether we are in the boot instant. If we are not in the boot instant, the nonterm: flags tell us that both parallel branches are selected. We only need two more bits to complete the representation: one for each parallel branch to define its internal state. Figure 9.7 gives two encoding examples.

Third example

The third example of tree encoding illustrates how representations of exclusive branches are multiplexed on the same bits to reduce the size of the state representation. Consider the following Esterel fragment, whose selection tree with its encoding are shown in Figure 9.8.

```
    pause
;
    [ await T || await S ]
;
```

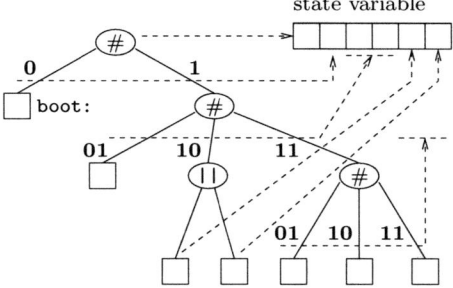

Figure 9.8: Bit allocation for the third small example

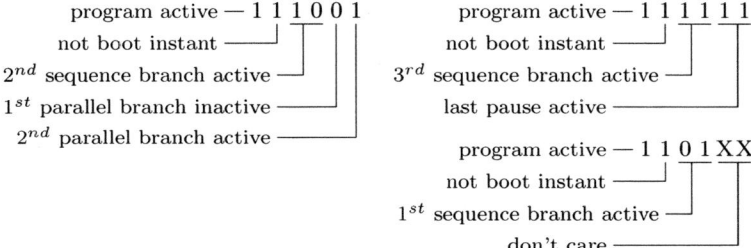

Figure 9.9: State encoding examples for the third small example

```
[ pause ; pause ; pause ]
```

Here, the last two bits of the representation are shared between the second and third branches of the sequence. Depending on the state of the program, these bits will be decoded by different flowgraph nodes. Figure 9.9 gives three encoding examples.

The encoding algorithm

The complexity of the state encoding algorithm is linear in the number of selection tree nodes. The algorithm performs a bottom-up traversal of the selection tree that incrementally computes the size of the representation for each of the selection sub-trees followed by a top-down traversal that actually assigns to selection nodes the integer constants (bit sequences) that will be used in the encoding of selection primitives.

For every program, the routine produces a bitwise representation pattern that is mapped on a sequence of actual C language variables. Table 9.1 gives statistics on the encoding algorithm run on some examples.

In addition to the number of state bits required, the table lists the number

Example	Circuit registers	Bits in C variables	Redundant selection bits
1-tcint	82	20	1
2-wristwatch	36	15	1
6-fuel	656	355	8
7-cabine	917	449	118
8-globalopt	1449	986	473

Table 9.1: State encoding results on some typical examples

of registers used to represent state in the circuit translation. This is a good measure of the redundancy of the circuit translation and, implicitly, of the circuit code. It also lists the number of bits used to represent branch selection, which could be optimized out as explained for the first example. For the two largest examples in particular, the implementation of this optimization may drastically improve the quality of the encoding.

9.2.2 Flowgraph Transformations

The encoding of the state representation on actual C variables determines which transformations are allowed. For each state access primitive, we must provide a sequence of C operations that perform the required task. The resulting code is a concurrent control-flow graph ready for scheduling.

The *test* primitives are the simplest to translate. They are only applied to parallel branch selection nodes that do not carry the **nonterm:** flag and are transformed into code that returns the status of the associated parallel branch.

Similarly, a *switch* primitive reads the associated choice bits from the appropriate state variable. To the calling **Switch** node, it returns the integer encoded on the bits, i.e., the index of the selected parallel branch. Thus, we are able to use `switch` C statements to dispatch control to the unique selected branch.

A *sync* primitive refers to a **parallel:** node with no **nonterm:** child. It tests the status of all the parallel branches and returns 1 (*true*) if at least one is selected. Otherwise it returns 0, informing the the calling **Sync** node that the parallel has terminated.

The most complex part of the primitive encoding concerns the state update primitives *enter* and *exit*, as most will generate no code. The reason for this is twofold: the logarithmic encoding of the choice makes most *exit* calls redundant: entering a branch of a sequence or test exits all the other branches as a side-effect, and in many cases successive *enter* calls can be "packed" into unique C variable assignments.

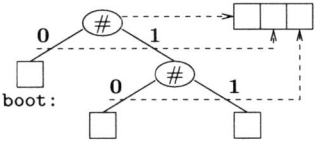

Figure 9.10: Bit allocation for pause ; pause

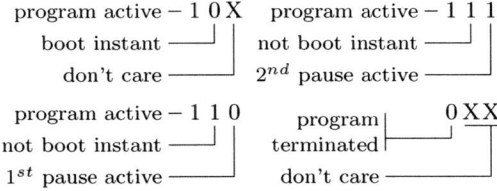

Figure 9.11: Possible states for pause ; pause and their encoding

A simple example

The following simple example illustrates the encoding steps.

    pause ; pause

Figure 9.10 shows the the selection tree and bit assignment for this example. Figure 9.11 lists the encoding of the four possible (macrostep) program states.

We are mainly interested in the associated GRC flowgraph, presented in Figure 9.12(a) and in the flowgraph obtained after state encoding, given in Figure 9.12(b). In the latter, the GRC state switch nodes have been replaced with state variable tests; the sequences of state update calls have each been replaced by single state variable assignments. Note the correspondence between the assign values and the codes of Figure 9.11.

Algorithm description

The only *exit* primitives that generate code are the ones corresponding to parallel branches that have a selection bit in the new encoding. The generated code sets the bit to 0.

Generating code for the *enter* primitives is more complex.

1. For a selection tree that has no parallel internal nodes, we only preserve the *enter* calls corresponding to the leaves. Each of them will perform an assignment of all the state variable bits, as in the previous example.

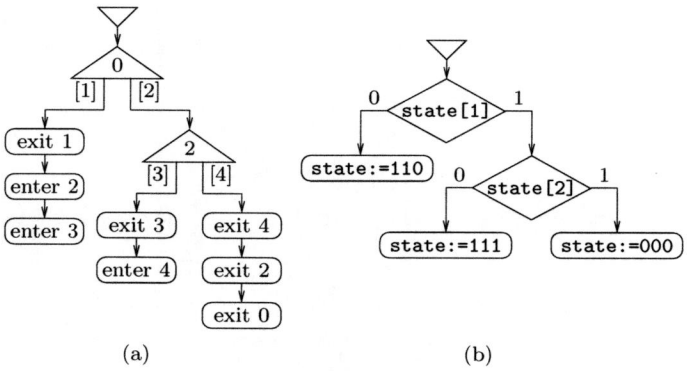

Figure 9.12: Flowgraph transformation due to state encoding

2. When a tree also contains parallel nodes, we reduce the problem to the previous case by partitioning the selection tree into disjoint sub-trees, each having the property that all its leaves are exclusive. On each of the sub-trees, we only keep *enter* calls corresponding to the leaves.

The partitioning algorithm produces sub-trees with the required property by cutting the arcs linking parallel nodes with their children.

- When a parallel node has no **nonterm:** child, then all arcs are deleted.
- When the parallel node has at least a **nonterm:** child, then all arcs are deleted save one, which must correspond to any of the **nonterm:** children.

The partitioning algorithm produces maximal sub-trees with the desired leaf exclusiveness property, which improves the quality of the state encoding.

Note that the partitioning process may have multiple solutions when a parallel node has multiple **nonterm:** children. One possible partition of the selection tree of **MainExample** is given in Figure 9.13. There is a second one, where the sub-tree rooted in node 0 contains the second parallel branch, not the first. Choosing which is best is not obvious.

9.2.3 Scheduling

We use a simple hierarchical greedy approach to statically schedule the flowgraph operators. Interleaving uses one auxiliary Boolean variable for every triggering condition. Figure 9.14 shows the C code generated for the GRC flowgraph of Figure 8.24, which corresponds to the **MainExample** program. We renamed variables to make their roles clearer.

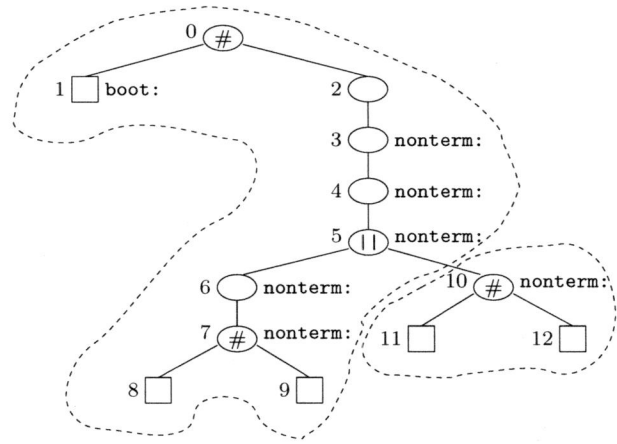

Figure 9.13: A partition of the selection tree of `MainExample`

The integer variable `state_0` holds the state representation. The state access and update code is separated for easier understanding and debugging. The internal signal variable `END_1` corresponds to the single copy (of incarnation index 1) of the internal signal `END` that is needed. The variable `SYNC_1` is set to 1 when trap `T` is exited. The two parallel branches are interleaved using the variable `CS_1`. The scope of all internal variables is determined at code generation time. In our case, no variable is declared/initialized in the boot instant or if the signal `SUSP` is present.

9.3 Code Generation for Cyclic Specifications

Certain GRC specifications remain cyclic after the application of our simplification techniques. In this class fall all the non-constructive Esterel programs and also the correct Esterel programs that, like the bus arbiter example of page 27, are intrinsically cyclic. Generating control-flow code for such specifications, for implementation or debugging purposes, involves two challenges: extending the techniques of the previous section with dynamic scheduling methods able to evaluate the cyclic parts of a specification; and at runtime, cheaply deciding when a program is non-constructive in a given context, to report it to the environment.

In this section, we suggest ways to generate fast control-flow code for syntactically correct Esterel program. We assume that expensive constructiveness check techniques, such as that developed by Toma [69], are impractical since they simply cannot handle large examples.

```
/* STATE VARIABLES, BOOT STATE */
unsigned long state_0 = 1u;

/* STATE ACCESS/UPDATE */
/* test the global program status */
bool test_0() { return state_0&1u; }

/* terminate the program */
void exit_0() { state_0&=~1u; }

/* test if the program is in the boot state */
unsigned long switch_expr_0() { return (state_0>>1)&1u; }

/* first parallel branch -- switch and state change */
unsigned long switch_expr_7() { return (state_0>>2)&1u; }
void enter_8() { state_0&=~6u; state_0|=2u; }
void enter_9() { state_0&=~6u; state_0|=6u; }

/* second parallel branch -- switch, and state change */
unsigned long switch_expr_10() { return (state_0>>3)&1u; }
void enter_11() { state_0&=~8u; state_0|=0u; }
void enter_12() { state_0&=~8u; state_0|=8u; }

/* THE REACTION FUNCTION */
int MainExample() {
  if (switch_expr_0()) {            /* boot instant test */
    if (!SUSP) {                    /* suspension test */
      /* auxiliary scheduling variables */
      bool SYNC_1 = 0; bool CS_1 = 0;
      /* local signal variables */
      bool END_1 = 0;

      /* 2nd parallel branch */
      if (switch_expr_10()) CS_1=1;  /* interleaving */
      else if (KILL) enter_12();

      /* 1st parallel branch */
      if (switch_expr_7()) { if (J) { END_1=1; enter_8(); } }
      else if (I) { EMIT_0(); enter_9(); }

      /* 2nd parallel branch, second part */
      if (CS_1) { if (END_1) SYNC_1=1; }

      /* parallel synchronization */
      if (SYNC_1) { exit_0(); }
    }
  } else { enter_8(); enter_11(); }
  __MainExample__reset_input();
  return test_0();
}
```

Figure 9.14: Code generated from the flowgraph of Figure 8.24, which corresponds to the MainExample program.

CODE GENERATION FOR CYCLIC SPECIFICATIONS 195

Constructive simulation is needed

When a GRC specification is cyclic, we do not have a guarantee of its correctness (constructiveness). Perhaps more important for code generation, no static schedule (topological ordering) of the flowgraph nodes guarantees the correct computation of reactions for all possible state and input event. In the absence of code duplication/resynthesis, the correct evaluation of reactions requires a constructive simulation of the flowgraph.

Reduction to SCCs

Unfortunately, practice shows that constructive simulation is an expensive process regardless of the domain in which it it performed (GRC, digital netlists, etc.). This is due to the necessity of dynamic node scheduling and to the use of the ternary logic operators we defined in Figure 3.3 instead of the Boolean operators implemented by most microprocessors.

Consequently, to generate efficient code, we need to be clever about performing constructive simulation. One obvious approach is to limit it to places where it is absolutely necessary, i.e., the strongly connected components (SCCs) of a specification. Nodes in an SCC have no static ordering without resynthesis or further dependency simplifications to remove static cycles.

Once the SCCs are identified, the cyclic flowgraph is transformed into an acyclic one by replacing each SCC with a unique flowgraph node representing its computation. The newly generated node, of a special "SCC" type, inherits all the dependencies relating SCC nodes to exterior ones. The new acyclic graph can now be scheduled using the techniques of the previous section, and we still have to define the constructive evaluation code needed to compute the behavior of the SCCs.

We illustrate how constructive evaluation code is embedded in the globally acyclic scheme with the following example.

```
signal S,T in
  pause ; present S then emit T end ;
  pause ; present T then emit S end
end
```

This code fragment is constructive. Figure 9.15 shows its selection tree and flowgraph. Note that the dependency simplifications of Section 8.4.2 would have removed the cycle, but we did not apply them for illustration purposes.

The cycle (in the dotted box) is formed by the signal test nodes. It does not disappear after circuit expansion (Figure 9.16), so the technique of Section 9.1 cannot be applied. We have to generate constructive evaluation code for the SCC.

When we isolate the cycle as an SCC node, we obtain the flowgraph of Figure 9.17. Applying the static scheduling technique of the previous section

196   CODE GENERATION FROM GRC

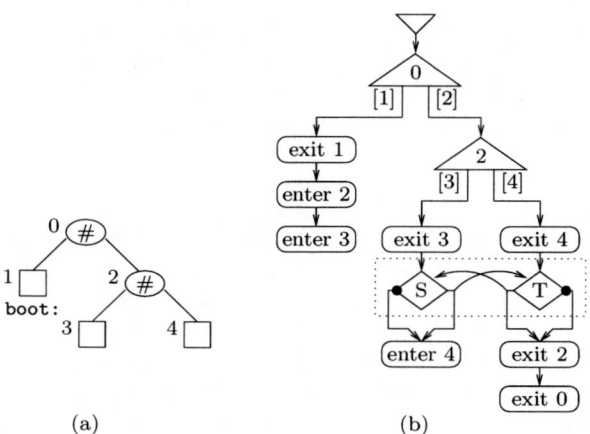

Figure 9.15: The GRC of the cyclic example

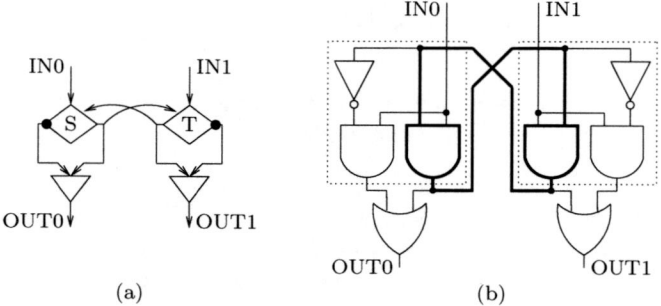

Figure 9.16: The SCC at GRC and circuit level

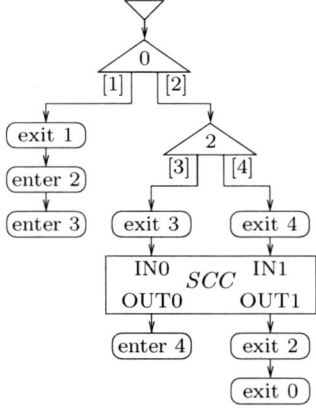

Figure 9.17: Globally acyclic graph for our cyclic example

gives the following.

```
if (switch_expr_0()) {              /* boot instant test */
  bool SCC_IN_0 = 0; bool SCC_IN_1 = 0;
  bool SCC_OUT_0 = 0; bool SCC_OUT_1 = 0;
  if (switch_expr_2()) SCC_IN_1=1;  /* second pause */
  else SCC_IN_0=1;                  /* first pause */

  /* SCC evaluation code goes here */

  if(SCC_OUT_0) enter_4();
  if(SCC_OUT_1) exit_0();
} else enter_3();
```

To complete the code we have to generate the dynamically-scheduled code associated with the SCC itself.
)

Code generation for SCCs—netlist simulation

The constructive semantics of GRC flowgraph nodes is given by circuit translation, so we perform constructive simulation of the SCC nodes at the netlist level. For example, the code generated for the SCC block of our small example is a simulator for the circuit of Figure 9.16(b).

The simplest approach to constructive netlist simulation is the one used in the INRIA compiler, with the "-I" option of the Esterel V5_92 circuit code generator.

The circuit associated with a GRC-level SCC is always combinational and corresponds to a single SCC in the circuit. To compute its behavior for given Boolean inputs, we may evaluate the gates that are not part of the circuit-level SCC with Boolean logic, but we must use three-valued constructive logic for the gates in the cycle.

To do this in software, we encode the wire values with Boolean variables. We use the usual dual-rail encoding: we associate two Boolean variables—w_true and w_false—to each wire w in the circuit-level SCC. The encoding is as follows.

w_true	w_false	w
0	0	$\bot$
1	0	*true*
0	1	*false*
1	1	-

The other wires are directly evaluated in the Boolean domain; we assign a single Boolean variable, with the same name, to each such wire. For clarity, we shall use the names w, w^i for SCC wires, and the names v, v^i for wires that do not belong to the SCC. The Boolean variables associated with v and v^i are denoted v and v_i.

Every three-value circuit gate becomes one or two assignment statements:

$w = w^1 \vee \ldots \vee w^n \vee v^1 \vee \ldots \vee v^m$
    w_true = w_1_true|...|w_n_true|v_1|...|v_m
    w_false= w_1_false&...&w_n_false&!v_1&...&!v_m
$v = w^1 \vee \ldots \vee w^n \vee v^1 \vee \ldots \vee v^m$
    v=w_1_true|...|w_n_true|v_1|...|v_m
$w = w^1 \wedge \ldots \wedge w^n \wedge v^1 \wedge \ldots \wedge v^m$
    w_true=w_1_true&...&w_n_true&v_1&...&v_m
    w_false=w_1_false|...|w_n_false|!v_1|...|!v_m
$v = w^1 \wedge \ldots \wedge w^n \wedge v^1 \wedge \ldots \wedge v^m$
    v=w_1_true&...&w_n_true&v_1&...&v_m
$w = \neg\, w^1$
    w_true=w_1_false;w_false=w_1_true;
$w = \neg\, v^1$
    w_true=!v_1;w_false=v_1;
$v = \neg\, w^1$
    v=w_1_false;
$v = \neg\, v^1$
    v=!v_1;

Wires that are not part of the circuit-level SCC are evaluated either before or after the evaluation of the SCC. The nodes in the SCC are evaluated by a stabilization loop, which repeatedly evaluates the equations in a fixed order until no wire value (variable) changes. Since constructive evaluation is monotonic, this process always terminates. In fact, since it is easy to bound the number of iterations required for this process to converge for a particular SCC, we can instead run each loop a fixed number of times computed at compile time. For instance, twice is enough for our two-gate cycle example.

```
/* SCC evaluation code */
bool v_0 = 0, v_1 = 0;
bool w_0_true = 0, w_0_false = 0;
bool w_1_true = 0, w_1_false = 0;
for (int cnt = 0 ; cnt<2 ; cnt++) {
  w_0_true = w_1_true & SCC_IN_0;
  w_0_false = w_1_false | !SCC_IN_0;
  w_1_true = w_0_true & SCC_IN_1;
  w_1_false = w_0_false | !SCC_IN_1;
}
if (!((w_0_true | w_0_false) & (w_1_true | w_1_false)))
  ERROR();
v_0 = w_1_false & SCC_IN_0; v_1 = w_0_false & SCC_IN_1;
SCC_OUT_0 = w_0_true | v_0; SCC_OUT_1 = w_1_true | v_1;
```

The line of code following the stabilization loop performs the run-time constructiveness check. If at least one SCC wire remains undefined (w_true|w_false==0) after two iterations of the evaluation loop, it calls an error-handling function. The constructiveness check can be safely removed if the program is proven to be constructive, e.g., using sccausal [69].

The number of iterations required to stablize an SCC depends on its structure and the wire evaluation order. As explained in Section 6.1, evaluating the body of the loop for a number of times equal to the the number of SCC gates is sufficient. However, this is overkill for large SCCs since the evaluation time will increase quadratically in the SCC size. Instead, more efficient iteration strategies can be used, such as the one proposed by Bourdoncle [16]. Such techniques order the gates in an SCC to improve the speed of their convergence.

Other approaches

Edwards [25, 53] proposes new approaches to generating simulation code for cyclic netlists. These use efficient resynthesis based on loop unrolling instead of the exact algorithms of Toma [69].

Lukoschus [47] proposes a source-to-source transformation that removes cycles by introducing new signals and duplicating code.

9.4 Benchmarks

This section describes the benchmarks we used to compare the `grc2c` compiler with other code generators. The testbench contains eight examples ranging from Berry's wristwatch to industrial-size examples of more than 13000 lines.

1-tcint	turbochannel bus controller
2-wristwatch	the digital watch model part of the INRIA distribution
3-atds100	video generator
4-mca200	shock absorber [20]
5-chorus	operating system model [72]
6-fuel	avionics fuel controller
7-cabine	avionics cockpit interface
8-global	avionics man-machine interface

The following table gives an idea of the specification size, initial (in LC and GRC form) and optimized (GRC), for the eight examples. It also gives the optimization time, measured on the test system—a PentiumIII/1GHz/128Mo running Linux.

Example	LC	GRC nodes		GRC, optimized			time
	nodes	init	sweep	nodes	%GRC	%LC	(s)
	(1)	(2)	(3)	(4)	(5)	(6)	(7)
1-tcint	516	924	798	375	46.9%	72.6%	1.06
2-ww	533	1025	834	366	43.8%	68.6%	1.04
3-atds100	1122	2307	2119	1059	49.9%	94.3%	2.30
4-mca200	3769	4897	4562	3596	78.8%	95.4%	6.58
5-chorus	4751	7385	6299	3539	56.1%	74.4%	7.90
6-fuel	4986	10449	8544	4516	52.8%	90.5%	10.20
7-cabine	10991	24359	18872	8037	42.5%	73.1%	23.10
8-global	15831	36525	18852	13253	70.3%	83.7%	55.59

The first column gives the LC statement counts produced by the frontend of the INRIA compiler for our eight examples. This is a good measure of the complexity of the initial Esterel specification and corresponds roughly the to number of source code lines. Column (2) gives the size (node count) of the GRC specifications that are obtained by control flow expansion*. Due to the inherent redundancy of the structural translation scheme the (large) figures are misleading. The third column gives a better estimate of the complexity of the initial GRC specification—the GRC node counts after a pass of selection-independent simplifications (similar to the *sweep* operation used in circuit optimizers). The next three columns give the size of the optimized

---
*Bear in mind the structure of the INRIA compiler, given in Figure 7.2.

GRC specification and compare it with the figures of the columns (1) and (3). The last column gives the time (in seconds user time) taken by the full optimization process for these examples.

The optimized size figures suggest the efficiency of our optimization technique may vary greatly from one example to the other. The optimization time figures show that these algorithms are indeed fast—apparently quasi-linear in the specification size.

Next, we compare the efficiency of the GRC-based control-flow code with the efficiency of code coming from four other code generators described in Chapter 7:

- the circuit-based compiler of the Esterel V5_92 system from INRIA (**Gates**);

- the same circuit-based compiler, when aggressive circuit optimizations are applied before C code generation using the SIS/blifopt script (**Gates_opt**);

- the **Synopsys** compiler developed by Stephen Edwards;

- the **Saxo-RT** compiler developed at France Telecom R&D.

The resulting C files were compiled using "gcc -O" and run for one million cycles on the same input event sequences.

The results are summarized in Tables 9.2 and 9.3. Missing figures are due to one of three cases: for the **Gates** code, they mean that gcc exhausted the system memory during compilation. For the **Gates_opt** code, they mean that SIS/blifopt was not able to optimize the circuit. For the **Synopsys** code, a missing figure simply means that the code was not available for tests. The best figures are emphasized for each example.

As expected, the non-optimized circuit code gives bad figures. For all our eight examples it is the slowest and biggest. The control-flow code generators (**Synopsys**, **Saxo-RT**, and **grc2c**) give similar figures in both speed and size. They are always faster than the circuit code, optimized or not. Among them, **grc2c** generates faster code. Both **Gates_opt** and **Synopsys** give good object file sizes for small examples. The differences in size between the control-flow code generators do not seem to be relevant.

Example	Gates	Gates_opt	Synopsys	Saxo-RT	grc2c
1-tcint	1.97	0.66	**0.25**	0.36	**0.25**
2-wristwatch	2.71	1.4	1.26	1.29	**0.95**
3-atds100	14.8	3.69	0.14	0.12	**0.08**
4-mca200	46.49	48.19	2.61	3.68	**1.88**
5-chorus	54.29	–	2.76	5.30	**1.10**
6-fuel	37.66	–	–	16.81	**15.65**
7-cabine	–	–	–	31.97	**29.26**
8-global	–	–	–	57.45	**43.27**

Table 9.2: Code speed (user time sec. for 1Mcycle)

Example	Gates	Gates_opt	Synopsys	Saxo-RT	grc2c
1-tcint	27.0	**11.4**	11.5	15.3	17.5
2-wristwatch	22.9	**11.2**	13.3	16.1	14.7
3-atds100	64.2	31.3	**21.7**	33.9	33.7
4-mca200	182.9	171.8	70.2	78.9	**67.7**
5-chorus	230.3	–	99.5	104.1	**98.6**
6-fuel	201.4	–	–	**147.5**	168.7
7-cabine	–	–	–	256.1	**196.8**
8-global	–	–	–	309.1	**273.9**

Table 9.3: Object code size, in Kb

10
The Columbia Compiler

One of the authors (Edwards) and his group have developed the Columbia Esterel Compiler at Columbia University starting in 2001. It is a separate code base from the INRIA compilers described in Section 7.3 and deliberately uses different compilation technology.

In this chapter, we describe two of the code-generation techniques in the Columbia Esterel Compiler (hereafter CEC). The first is a clever dynamic technique that manipulates very efficient linked lists of code fragments that need to be executed. In this sense, it resembles the Saxo-RT compiler from France Telecom R&D, but tries to reduce the asymptotic running time by only doing work for code that actually runs.

The second technique blasts apart the GRC representation of an Esterel program and then re-forms it in a program dependence graph (PDG) [31], a very abstract form of a concurrent control-flow graph that exposes even more concurrency than the GRC representation. Such granularity allows the compiler to restructure the code in such a way to reduce the amount of context-switching-related overhead and produce even faster code. The main novelties here are the application of the the program dependence graph (PDG) formalism to an Esterel context and the algorithm, presented later, for generated sequential code from the concurrent PDG.

10.1 The Dynamic Technique

This technique, originally presented at the SLAP workshop in 2004 [28], was inspired in part by Maurer's Event-Driven Conditional-Free paradigm [50], although his implementation is geared to logic network simulation and does not appear applicable to Esterel . Interestingly, he writes his examples using a C-like notation that resembles the gcc computed *goto* extension used in CEC, but apparently he uses inlined assembly instead of the gcc extension.

The technique also resembles that in the Synopsys compiler [26], but this technique uses a more dynamic scheduler. During a cycle, the Synopsys compiler maintains a set of state variables, one per running thread. At each context switch point, the compiler generates code that performs a multiway branch on one of these variables. While this structure is easier for the compiler to analyze, it is not able to quickly skip over as much code as the dynamic technique presented here.

Esterel's semantics require any implementation to deal with three issues: the concurrent execution of sequential threads of control within a cycle, the scheduling constraints among these threads due to communication dependencies, and how (control) state is updated between cycles.

The dynamic technique generates C code that executes concurrently-running threads by dispatching small groups of instructions that can run without a context switch. These blocks are dispatched by a scheduler that uses linked lists of pointers to code that will be executed in the current cycle. The scheduling constraints are analyzed completely by the compiler before the program runs and affects both how the Esterel programs are divided into blocks and the order in which the blocks may execute. Control state is held between cycles in a collection of variables encoded with small integers.

10.1.1 An Example

We illustrate the operation of the CEC compiler on the small Esterel program in Figure 10.1. It models a shared resource using three groups of concurrently-running statements. The first group (*await I* through *emit O*) takes a request from the environment on signal I and passes it to the second group of statements (*loop* through *end loop*) on signal R. The second group responds to requests on R with the signal A in alternate cycles. The third group simply makes Q a delayed version of R.

This simple example illustrates many challenging aspects of compiling Esterel. For example, the first thread communicates with and responds to the second thread in the same cycle, i.e., the presence of R is instantaneously broadcast to the second thread, which, if the *present* statement is running, observes R and immediately emits A in response. In the same cycle, emitting A causes the *weak abort* statement to terminate and send control to *emit O*.

As is often the case, the inter-thread communication in this example means that it is impossible to execute the statements in the first thread without interruption: those in the second thread may have to execute partway through. Ensuring the code in the two threads executes in the correct, interleaved order at runtime is the main compilation challenge.

CEC translates Esterel into a variant of the GRC format called GRC_{CEC}. A GRC_{CEC} representation, like GRC, consists of a selection tree that represents the state structure of the program and an acyclic concurrent control-flow graph that represents the behavior of the program in each cycle. We invite the user to compare it with the definitions of Chapter 8.

```
module Example:
input I, S;
output O, Q;

signal R, A in
  every S do
        await I;
        weak abort
          sustain R
        when immediate A;
        emit O
    ||
      loop
        pause;
        pause;
        present R then
          emit A
        end present
      end loop
    ||
      loop
        present R then
          pause;
          emit Q
        else
          pause
        end present
      end loop
  end every
end signal

end module
```

Figure 10.1: A simple Esterel module modeling a shared resource. Here, the first thread waits for I and responds by sustaining R (request) until A (acknowledge) is returned by the second thread. The third thread makes Q a delayed version of R.

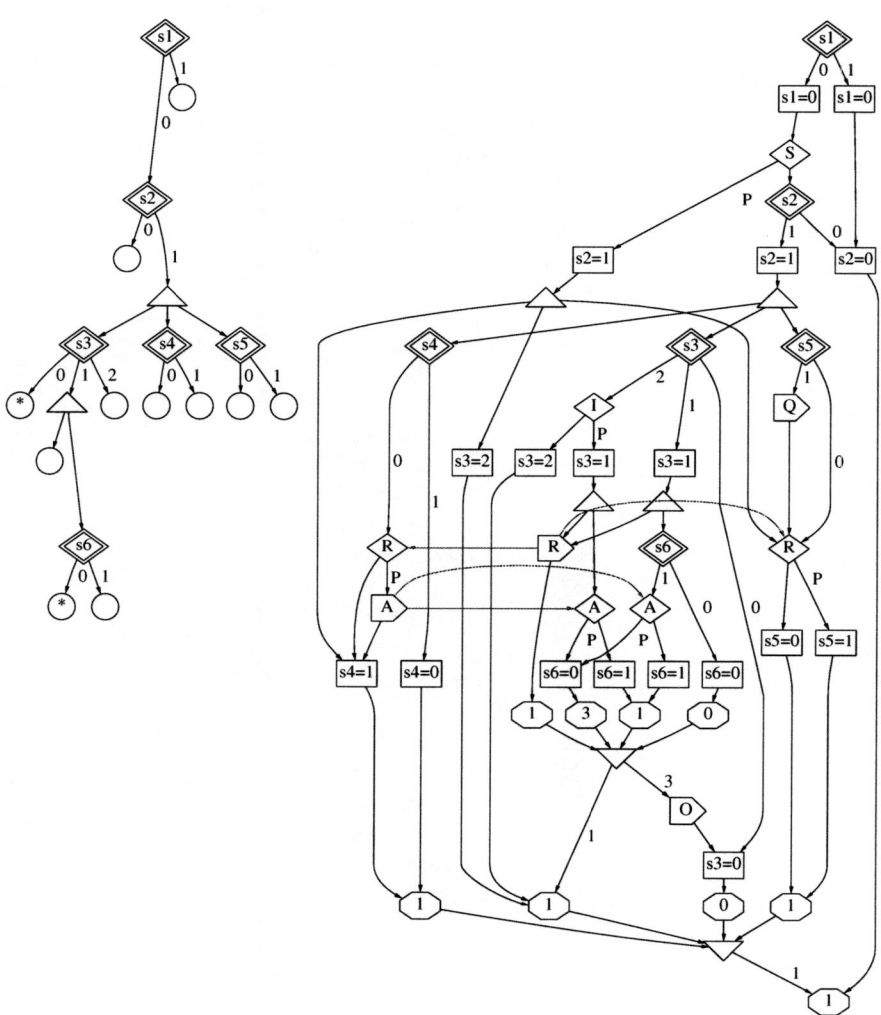

Figure 10.2: The (simplified) GRC$_{CEC}$ graph for the program in Figure 10.1. It consists of a selection tree (left) and a control-flow graph (right).

The selection tree

The selection tree (left of Figure 10.2) is the simpler half of a GRC$_{\text{CEC}}$ representation. The tree consists of three types of nodes: leaves (circles) that represent atomic states, e.g., *pause* statements; exclusive nodes (double diamonds) that represent choice, i.e., if an exclusive node is active, exactly one of its subtrees is active; and fork nodes (triangles) that represent concurrency, i.e., if a fork node is active, all of its subtrees are active.

Although the selection tree is used during the optimization phase of CEC, for the purposes of code generation it is just a complicated way to enumerate the variables needed to hold the control state of an Esterel program between cycles. Specifically, each exclusive node represents an integer-valued variable that stores which of its children may be active in the next cycle. In Figure 10.2, these are labeled s1 through s6. We encode these variables in the obvious way: 0 represents the first child, 1 represents the second, and so forth.

The control-flow graph

The control-flow graph (right of Figure 10.2) is a much richer object and the main focus of the code-generation procedure. It is a traditional flowchart consisting of actions (rectangles and pointed rectangles, indicating signal emission) and decisions (diamonds) augmented with fork (triangles), join (inverted triangles), and terminate (octagons) nodes.

The control-flow graph is executed once from entry to exit for each cycle of the Esterel program. The nodes in the graph test and set the state variables represented by the exclusive nodes in the selection tree and test and set Boolean variables that represent the presence/absence of signals.

The fork, join, and terminate nodes are responsible for Esterel's concurrency and trap constructs. When control reaches a fork node, it is passed to all of the node's successors. Such separate threads of control then wait at the corresponding join node until all the incoming threads have arrived.

Esterel's structure induces properly nested forks and joins. Specifically, each fork has exactly one matching join, control does not pass among threads before the join, and control always reaches the join of an inner fork before reaching a join of an outer fork. In Figure 10.2, each join node has two corresponding forks, and the topmost two forks are owned by the lowest join.

Together, join nodes—the inverted triangles in Figure 10.2—and their predecessors, terminate nodes—the octagons—implement two aspects of Esterel's semantics: the "wait for all threads to terminate" behavior of concurrent statements and the "winner-take-all" behavior of simultaneously-thrown traps. Each terminate node is labeled with its integer completion code, as defined in Section 3.3.4. Once every thread in a group started by a fork has reached the corresponding join, control passes from the join along its outgoing arc labeled with the highest completion code of all the threads.

Consider the behavior of the program in Figure 10.1 represented by the control-flow graph on the right of Figure 10.2. The topmost node tests state variable s1, which is initially set to 1 to indicate the program has not yet started. The test of S immediately below the nodes that assign 0 to s1 implements the *every S* statement by restarting the two threads when S is present (indicated by the label P on the arc from the test of S). The test of s2 just below S encodes whether the body of the *every* has started and should be allowed to proceed.

The fork just below the rightmost s2=1 action resumes the three concurrent statements by sending control to the tests of state variables s3, s4, and s5. Variable s3 indicates whether the first thread is at the *await I* (=2), sustaining R while checking for A (=1), or has terminated (=0). Variable s6 could actually be removed. It is a side effect arising from how our compiler translates the *weak abort* statement into two concurrent statements, one of which tests A. The variable s6 indicates whether the statement that tests A has terminated, something that can never happen.

When s3 is 1 or s3 is 2 and I is present, these two threads emit R and test A. If A is present, control passes through the terminate 3 node to the inner join. Because this is the highest exit level (the other thread, which emits R, always terminates at level 1), this causes control to pass from the join along the arc labeled 3 to the node for *emit O* and to the action s3=0, which effectively terminates this thread.

The second thread, topped by the test of s4, either checks R and emits A in response, or simply sets s4 to 0 so it will be checked in the next cycle.

The third thread, which starts at the test of s5, initially emits Q if s5 is 1, then sets s5 to 1 if R is present.

Although the behavior of the state assignments, tests, and completion codes is fairly complicated, it is easy to translate into imperative code. Unfortunately, concurrency complicates things: because two of the threads cannot be executed atomically since the presence of signals R and A must be exchanged during their execution within a cycle. Generating sequential code that implements this concurrency is the main trick in the CEC dynamic code generation technique.

Differences between GRC *and* GRC_{CEC}

Although the semantics of the two GRC variants are identical and their overall structure is the same, there are several subtle differences. The most visible is the replacement of **Sync** GRC nodes with the combination of terminators and join nodes in GRC_{CEC}. This implies, for instance, that a GRC_{CEC} flowgraph cannot be translated into a circuit by replacing each operator with a small circuit independently of the other nodes. Specifically, a join node must be translated together with the associated terminators.

Maybe more important, GRC_{CEC} makes more encoding choices than GRC, explicitly using a logarithmic encoding of state variables. Thus, a GRC_{CEC} representation can be seen as intermediate between its GRC

THE DYNAMIC TECHNIQUE   209

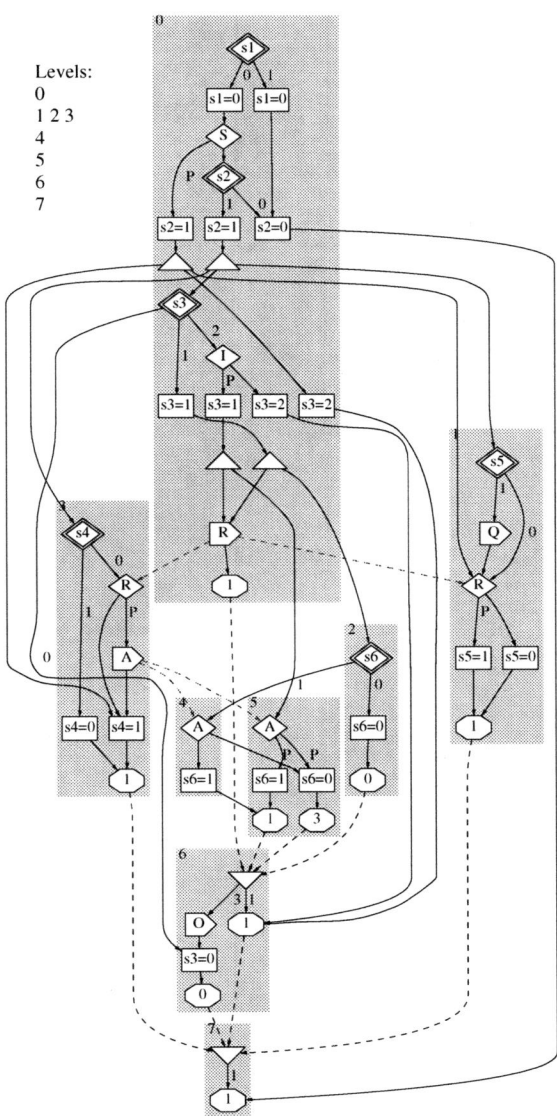

Figure 10.3: The control-flow graph from Figure 10.2 divided into blocks. Control arcs reaching join nodes have been replaced with (dashed) data dependencies to guarantee each block has at most one active incoming control arc.

```
#define sched1a next1 = head1, head1 = &&C1a
#define sched1b next1 = head1, head1 = &&C1b
#define sched2  next2 = head1, head1 = &&C2
#define sched3a next3 = head1, head1 = &&C3a
#define sched3b next3 = head1, head1 = &&C3b
#define sched4  next4 = head2, head2 = &&C4
#define sched5a next5 = head3, head3 = &&C5a
#define sched5b next5 = head3, head3 = &&C5b
#define sched5c next5 = head3, head3 = &&C5c
#define sched6a next6 = head4, head4 = &&C6a
#define sched6b next6 = head4, head4 = &&C6b
#define sched6c next6 = head4, head4 = &&C6c
#define sched7a next7 = head5, head5 = &&C7a
#define sched7b next7 = head5, head5 = &&C7b

int cycle() {
  void *next1;
  void *next2;
  void *next3;
  /* other next pointers */

  void *head1 = &&END_LEVEL_1;
  void *head2 = &&END_LEVEL_2;
  /* other level pointers */

  if (s1) { s1 = 0; goto N26; }
  else {
    s1 = 0;
    if (S) {
      s2 = 1; code0 = -1;
      sched7a; sched1b; sched3b;
      s3 = 2; sched6b;
    } else {
      if (s2) {
        s2 = 1; code0 = -1; sched7a; sched1a; sched3a;
        switch (s3) {
        case 0: sched6c; break;
        case 1: s3 = 1; code1 = -1; sched6a; sched2; goto N38;
        case 2:
          if (I) {
            s3 = 1; code1 = -1; sched6a; sched5a;
N38:        R = 1; code1 &= -(1 << 1);
          } else { s3 = 2; sched6b; }
          break;
        } } else {
N26:    s2 = 0; sched7b;
      } } }
  goto *head1;

C1a: if (s5) Q = 1;
C1b: if (R) s5 = 1;
     else s5 = 0;
     code0 &= -(1 << 1);
     goto *next1;
C2:  if (s6) sched4;
     else s6 = 0;
     goto *next2;
C3a: if (s4) s4 = 0;
     else {
       if (R) A = 1;
C3b:   s4 = 1;
     }
     code0 &= -(1 << 1);
     goto *next3;
END_LEVEL1: goto *head2;
```

Figure 10.4: The code CEC generates for the first two levels of Figure 10.3: clusters 0, 1, 2, and 3 (reformatted to fit space).

counterpart and the result of the state encoding phase in the INRIA control-flow code generator (cf. Section 9.2 and Figure 9.12).

While not a part of the GRC format itself, the reincarnation scheme used in the CEC and INRIA control-flow code generators are different. CEC attempts to minimize code duplication. This is why two fork nodes correspond to a single join node in our small example.

10.1.2 Sequential Code Generation

The dynamic code generation technique relies on the following observations: while arbitrary clusters of nodes in the control-flow graph cannot be executed without interruption, many large clusters often can be; these clusters can be chosen so that each is invoked by at most one of its incoming control arcs; because of concurrency, a cluster's successors may have to run after some intervening clusters have run; and groups of clusters without any mutual data or control dependency can be invoked in any order (i.e., clusters are partially ordered).

The key trick comes from this last observation: because the clusters within a level can be invoked in any order, it suffices to use an inexpensive singly-linked list to track which clusters must be executed in each level. By contrast, most discrete-event simulators [2] are forced to use a more costly data structure such as a priority queue for scheduling.

The overhead in this scheme approaches a constant amount per cluster *executed*. By contrast, the overhead of the Saxo-RT compiler is proportional to the total number of clusters in the program, regardless of how many actually execute in each cycle, and the overhead in the netlist compilers is even higher: proportional to the number of statements in the program.

CEC divides a concurrent control-flow graph into clusters of nodes that can execute atomically and orders these clusters into levels that can be executed in any order. The generated code contains a linked list for each level that stores which clusters need to be executed in the current cycle. The code for each cluster usually includes code for scheduling a cluster in a later level: a simple insertion into a singly-linked list.

Figure 10.3 shows the effect of running the clustering algorithm on the control-flow graph of Figure 10.2. The algorithm identified eight clusters, but this is no ideal: a better algorithm would have combined clusters 4 and 5, but it is not surprising that our simple-minded algorithm misses the optimum since the optimum scheduling problem is NP-complete (see Edwards [27]).

After eight clusters were identified, our levelizing algorithm, which uses a simple relaxation technique, grouped them into the six levels listed at the top of Figure 10.3. It observed that clusters 1, 2, and 3 have no interdependencies, can be executed in any order, and placed them together in the second level. The other clusters are all interdependent and must be executed in the order identified by the levelizing algorithm.

The main trick is the semi-dynamic scheduler based on a sequence of linked lists. The generated code maintains a linked list of entry points for each level. In Figure 10.4, the `head1` variable points to the head of the linked list for the first level (the complete code has more such variables) and the `next1` through `next3` variables point to the successors of clusters 1 through 3.

The code in Figure 10.4 takes advantage of gcc's computed *goto* extension to C. This makes it possible to take the address of a label, store it in a void pointer (e.g., `head1 = &&C1a`) and later branch to it (e.g., `goto *head1`) provided this does not cross a function boundary. While not strictly necessary (in fact, we include a compiler flag that changes the generated code to use *switch* statements embedded in loops instead of *gotos*), using this extension substantially reduces scheduling overhead since a typical switch statement requires at least two bounds checks plus either a jump table lookup or a cascade of conditionals.

Figure 10.5 illustrates the behavior of these linked lists. Figure 10.5(a) shows the condition at the beginning of every cycle: every level's list is empty—the *head* pointer for each level points to its *END_LEVEL* block. If no blocks where scheduled, the program would execute the code for cluster 0 only.

Figure 10.5(b) shows the pointers after executing *sched3a*, *sched1b*, and *sched4* (note: this particular combination cannot occur in practice). Invoking the *sched3a* macro (see Figure 10.4) inserts cluster 3 into the first level's linked list by setting next3 to the old value of head1—END_LEVEL1—and setting head1 to point to C3a. Invoking *sched1b* is similar: it sets next1 to the new value of head1—C3a—and sets head1 to C1b. Finally, invoking *sched4* inserts cluster 4 into the linked list for the second level by setting next4 to the old value of head2—END_LEVEL2—and setting head2 to C4. This series of scheduling steps produces the arrangement of pointers shown in Figure 10.5(b).

Because clusters in the same level may be executed in any order, clusters in the same level can be scheduled cheaply by inserting them at the beginning of the linked list. The `sched` macros do exactly this. Note that the level of each cluster is hardwired since this information is known at compile time.

A powerful invariant that arises from the structure of the control-flow graph is the guarantee that each cluster can be scheduled at most once during any cycle. This makes it unnecessary for the generated code to check that it never inserts a cluster in a particular level's list more than once.

As is often the case, both cluster 1 and 3 have multiple entry points. This is easy to support because the structure of the graph guarantees that at most one entry point for each cluster will be be scheduled each cycle.

CEC uses the dominator-based code structuring algorithm described in Edwards [27] to generate structured code for each cluster. Some *goto*s are necessary to avoid duplicating code. Figure 10.4 has two: `N26` and `N38`.

THE DYNAMIC TECHNIQUE 213

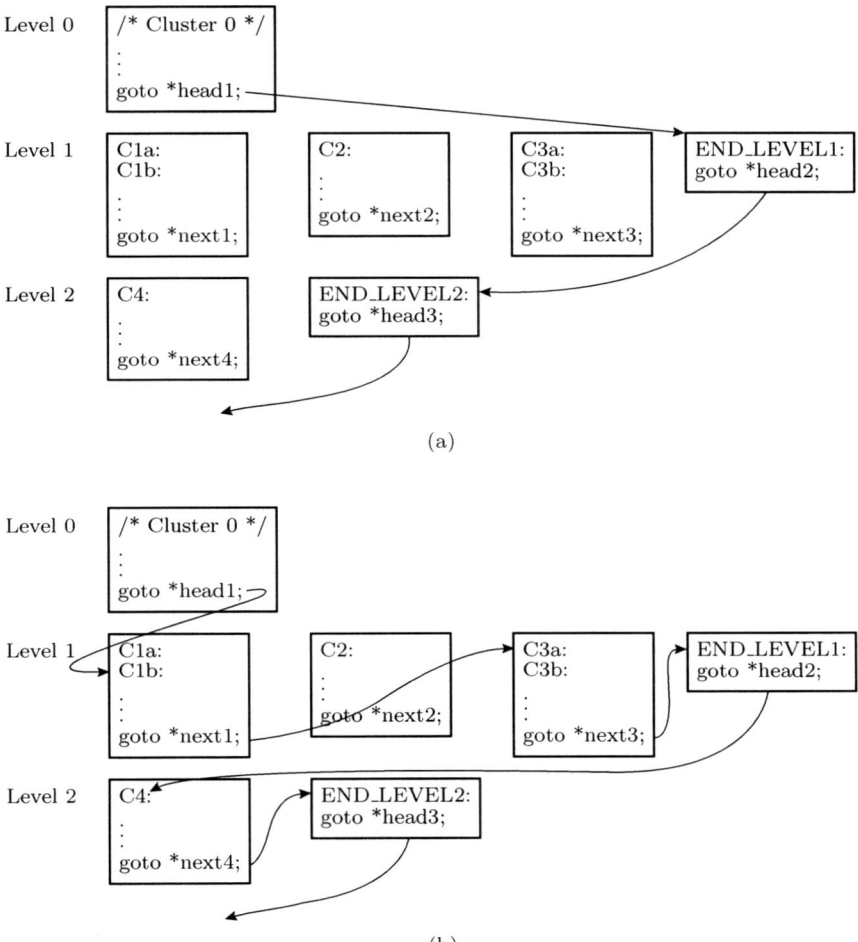

Figure 10.5: Cluster code and the linked list pointers. (a) At the beginning of a cycle. (b) After executing sched3a, sched1b, and sched4.

1: add the topmost control-flow graph node to F, the frontier set
2: **while** F is not empty **do**
3:    randomly select and remove f from F
4:    create a new, empty pending set P
5:    add f to P
6:    set C_i to the empty cluster
7:    **while** P is not empty **do**
8:      randomly select and remove p from P
9:      **if** p is not clustered and all of p's predecessors are **then**
10:         add p to C_i (i.e., cluster p)
11:         **if** p is not a fork node **then**
12:            add all of p's control successors to P
13:         **else**
14:            add the first of p's control successors to P
15:         add all of p's successors to F
16:      remove p from F
17:    **if** C_i is not empty **then**
18:      $i = i + 1$ (move to the next cluster)

Figure 10.6: The clustering algorithm. This takes a control-flow graph with information about control and data predecessors and successors and produces a set of clusters $\{C_i\}$, each of which is a set of nodes that can be executed without interruption.

10.1.3 The Clustering Algorithm

Figure 10.6 shows our clustering algorithm. It is heuristic and certainly could be improved, but is correct and produces reasonable results.

One important modification is made to the control-flow graph before our clustering algorithm runs: all control arcs leading to join nodes are removed and replaced with data dependency arcs, and a control arc is added from each *fork* to its corresponding *join*. This guarantees that no node ever has more than one active incoming control arc (before this change, each *join* had one active incoming arc for every thread it was synchronizing). Figure 10.3 partially reflects this restructuring: the additional arcs off the forks have been omitted to simplify an already complex diagram. This transformation also simplifies the clustering algorithm, which would otherwise have to handle *join*s specially.

The algorithm manipulates two sets of CFG nodes. The frontier set F holds the set of nodes that might start a new cluster, i.e., those nodes with at least one clustered predecessor. F is initialized in line 1 with the first node that can run—the entry node for the control-flow graph—and is updated in line 15 when the node p is clustered. The pending set P, used by the inner loop in lines 7–16, contains nodes that could be added to the existing cluster. P is initialized in line 5 and updated in lines 12–14.

The algorithm consists of two nested loops. The outermost (lines 2–18) selects a node f at random from the frontier F (line 3) and tries to start a cluster around it by adding it to the pending set P (line 5). The innermost (lines 7–16) selects a node p at random from the pending set P (line 8) and tries to add it to the current cluster C_i.

The test of p's predecessors in line 9 is key. It ensures that when a node p is added to the current cluster, all its predecessors have already been clustered. This ensures that in the final program, all of p's predecessors will be executed before p. If this test succeeds, p is added to the cluster under construction in line 10.

All of p's control successors are added to the pending set in line 12 if p is not a fork node, and only the first if p is a fork (line 14). This test partially breaks clusters at fork nodes, ensuring that all the nodes within a cluster are connected with sequential control flow, i.e., they do not run concurrently. Always choosing the first successor under a fork is arbitrary and may not be the best. In general, the optimum choice of which thread to execute depends on the entire structure of the threads. But even the simple-minded rule of always executing the first thread under a fork, as opposed to simply scheduling it, greatly reduces the number of clusters and significantly improves performance.

10.2 The Program Dependence Graph Approach

The second code generation technique we describe from the Columbia Esterel Compiler translates a GRC_{CEC} representation into the well-known program dependence graph (PDG) representation [31], then generates efficient code for it. While some PDG-to-sequential-code algorithms already existed before CEC, the novelty of our algorithm is that it can translate all legal PDGs, not just those that could be translated into sequential code without adding additional predicates. This technique was first published in 2004 [73].

CEC first performs a syntax-directed translation of an Esterel program into the GRC_{CEC} representation. It then converts this into a PDG using a slight modification of the algorithm due to Cytron et al. [23] to handle Esterel's concurrent constructs.

The main advance here is an algorithm that restructures a program dependence graph with arbitrary acyclic data dependencies into one that has a direct translation into sequential code. Unlike a PDG generated from purely sequential code, it is not usually possible to translate the PDG produced from Esterel directly into sequential code because communication patterns in the Esterel program may force concurrently-running threads to be interleaved. This can be solved by either duplicating code, a potentially costly operation that may produce an exponential increase in code size, or by inserting additional guard variables and predicates. CEC takes the second approach, using heuristics to choose where to cut the PDG and introduce predicates, and produce a semantically equivalent PDG that does have a simple sequential representation. CEC uses a modified version of Simons and Ferrante's algorithm [67] to produce a sequential control-flow graph from this restructured PDG and finally generate sequential C code from it.

Ferrante and Mace [30] were the first to propose an algorithm for generating sequential code from an acyclic PDG, but their technique only works when no node duplication (or equivalently, the addition of predicates) is necessary. Later, Simons and Ferrante [67] presented an efficient algorithm for generating sequential code from an acyclic PDG. Their major contribution is a technique for computing "external edge" information for each node and using this during the synthesis procedure. The input to their algorithm is limited to a graph with only control dependencies; they assume data dependencies have somehow been incorporated into the control dependencies.

Building on Simons and Ferrante's work, Steensgaard [68] removed the requirement that the control dependencies in the PDG be acyclic, thereby allowing loops in the generated code (earlier work assumed that loops had somehow been removed), but still assumed that the generated code did not require either node duplication or the insertion of additional predicates. CEC does not use Steensgaard's cyclic extensions because they are unnecessary for Esterel.

This technique extends Simons and Ferrante's in two ways. First, it uses a cutting algorithm that restructures the PDG and inserts additional predicate nodes before it is passed to Simons and Ferrante's basic algorithm, making it work for all valid acyclic PDGs. Second, it considers data dependencies to generate correct code for all valid PDGs.

The algorithm works in three phases (see Figure 10.8). First, it compute a schedule—a total order of all the nodes in the PDG (Section 10.2.2). This procedure is exact in the sense that it always produces a correct result, but heuristic in the sense that it may not produce an optimal result. Second, this schedule is used to guide a procedure for restructuring the PDG that slices away parts of the PDG, moves them elsewhere, and inserts assignments and tests of guard variables to preserve the semantics of the PDG (Section 10.2.3). Finally, CEC uses a slightly enhanced version of the sequentializing algorithm due to Simons and Ferrante to produce a control-flow graph (Section 10.2.4). Unlike Simons and Ferrante's algorithm, the variant used here always completes because of the restructuring phase. The experimental results presented in Section 10.3 show this technique can produce code that runs as much as thirty times faster than the (reference) circuit code generated by the INRIA compiler.

10.2.1 Program Dependence Graphs

This code generator models the Esterel program using a variant of Ferrante, Ottenstein and Warren's [31] program dependence graph. The PDG for a program is a directed graph whose nodes represent statements and whose arcs represent the partial ordering among statements that must be followed

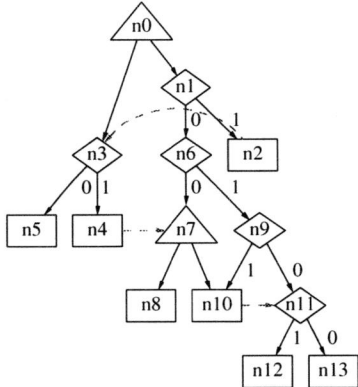

Figure 10.7: A program dependence graph requiring interleaving. Diamonds are predicate nodes, triangles are forks, and rectangles are statements. Solid lines are control arcs; dashed lines are data.

to preserve the program's semantics. In some sense, the PDG removes the maximum number of dependencies among statements without changing the program's meaning.

A PDG is a rooted, directed acyclic graph $G = (S, P, F, r, c, D)$. S, P, and F are disjoint sets of statement, predicate, and fork nodes. Together, these form the set of all vertices in the graph, $V = S \cup P \cup F$. $r \in V$ is the distinguished root node. $c : V \to V^*$ is a function that returns the vector of control successors for each node (i.e., they are ordered). Each vertex may have a different number of successors. $D \subset V \times V$ is a set of data edges. If $c(v_1) = (v_2, v_3, v_4)$, then node v_1 can pass control to v_2, v_3, and v_4. The set of control edges can be defined as $C = \{(m, n) : c(m) = (\ldots, n, \ldots)\}$, i.e., (m, n) is a control edge if n is some element of the vector $c(m)$. If a data edge $(m, n) \in D$, then m can pass data to node n.

The semantics of the graph rely mostly on the vertex types. A statement node $s \in S$ is the simplest: it represents a computation with a side-effect (e.g., assigning a value to a variable) and has no outgoing control arcs. A predicate node $p \in P$ also represents a computation but has outgoing control arcs. When executed, a predicate arc passes control to exactly one of its control successors depending on the outcome of the computation it represents. A fork node $f \in F$ does not represent computation; instead it merely passes control to all of its control successors. We call them fork nodes to emphasize that they represent concurrency; other authors call them "region nodes," although they mean the same thing.

In addition to being rooted and acyclic, the structure of the directed graph (V, C) satisfies two important constraints.

The predicate least common ancestor rule (PLCA) requires that for any node $n \in V$ with two different control paths to it from the root, the least common ancestor (LCA) of any pair of distinct predecessors of n is a predicate node. PLCA ensures that there is at most one active path to any node. If the LCA node was a fork, control could conceivably follow two paths to n, implying multiple executions of the same node, something we explicitly wish to prohibit.

The no post-dominance rule: if n is a descendant of a node m then there is some path from m to some statement node that does not include n. This is because we insist that the PDG has eliminated unnecessary control dependencies among nodes. Otherwise, m and n would have been placed under a common fork.

10.2.2 Scheduling

Building a sequential control-flow graph from a program dependence graph requires ordering the concurrently-running nodes in the PDG. In particular, the children of each fork node are semantically concurrent but must be executed in some sequential order. The main challenge is dealing with cases where data dependencies among children of a fork force their execution to be interleaved.

Figure 10.7 shows a PDG that illustrates the challenge. In this graph, data dependencies require n3 to be executed after n2 and n7 to be executed after n4. Thus, the two subtrees under node n0 cannot be executed one after the other; they must be interleaved. The generated code must ensure nodes n2, n3, n4, and n7 execute in that order. This example is fairly straightforward, but such interleaving can become very complicated in large graphs with lots of data dependencies and reconverging control-flow such as that at node n10.

Duplicating certain nodes in the PDG of Figure 10.7 could produce a semantically equivalent graph with no interleaving but it also could cause an exponential increase in graph size. Instead, we restructure the graph and add predicates that test guard variables. Unlike node duplication, this introduces extra runtime overhead, but it can produce much more compact code.

The CEC approach inserts guard variable assignments and tests based on cuts implied by a topological ordering of the nodes in a PDG. A cut represents a switch from an incompletely-scheduled child of a fork to another child of the same fork. It divides the nodes under a branch of a fork into two or more subgraphs.

To minimize the runtime overhead introduced by this technique, CEC tries to add few guard variables by making as few cuts as possible. Ferrante, Mace, and Simons [30] showed this minimum cut problem is NP-complete, so CEC attempts to solve it cheaply with heuristics.

CEC first computes a schedule for the PDG then follows this schedule to find cuts where interleavings occur. It uses a heuristic to choose a good schedule, i.e., one implying few cuts, that tries to choose a good order in which

procedure Main
  Clear the visited set
  PriorityDFS(root node of G)
  Clear the schedule and visited set
  ScheduleDFS(root node of G)
  Restructure()
  Fuse guard variables
  Generate sequential code from G'

Figure 10.8: The Main procedure

to visit each node's successors. CEC identifies the cuts while restructuring the graph.

Ordering node successors

To improve the quality of the generated cuts, CEC uses the heuristic algorithm in Figure 10.9 to influence the scheduling algorithm. It computes an order for successors of each node that the DFS-based scheduling procedure in Figure 10.10 uses to visit the successors.

CEC assigns each successor a priority vector of three integers (p_1, p_2, p_3) computed using the procedure described below, and later visit the successors in descending priority order while constructing the schedule. Priority vectors are totally ordered: $(p_1, p_2, p_3) > (q_1, q_2, q_3)$ if $p_1 > q_1$, or $p_1 = q_1$ and $p_2 > q_2$, or if $p_1 = q_1$, $p_2 = q_2$, and $p_3 > q_3$. For each node n, the A array holds the set of nodes at or below n that have any incoming or outgoing data arcs.

The first priority number of s_i (the ith subgraph under node n) counts the number of incoming data dependencies. It is the number of incoming data arcs minus the number of outgoing data arcs to/from any other subgraphs under node n.

The second priority number counts the number of elements that "pass through" the subgraph s_i. Specifically, it decreases by one for each incoming data arcs from a subgraph s_j to a node in s_i with a node m that is a descendant of s_i that has an outgoing data arc to another subgraph s_k ($j \neq i$ and $k \neq i$, but k may equal j).

The third priority counts incoming and outgoing data arcs connected to any nodes in sibling subgraphs. It is the total number of incoming data arcs minus the number of outgoing data arcs.

Finally, a node without any data arc entering or leaving its descendants is assigned a minimum first priority number.

Under these definitions, the priority of the left branch under n0 in Figure 10.7 is $(0, -1, 0)$, and that the right branch is $(0, 0, 0)$. Arcs from n2 to n3 and from n4 to n7 both affect the first priority number, but their effects cancel

procedure PriorityDFS(n)
   if n has not been visited **then**
     add n to the visited set
     for each control successor s of n **do**
       PriorityDFS(s)
       $A[n] = A[n] \cup A[s]$
     for each control successor s of n **do**
       ComputeSuccPriority(n, s)
     if n has any incoming or outgoing data arcs **then**
       add n to $A[n]$

procedure ComputeSuccPriority(n, s)
  $(a, b, c) = (0, 0, 0)$ {initialize priorities}
  if s has neither incoming nor outgoing data arcs **then**
    a = minimum priority number
    return
  for each $j \in A[s]$ **do**
    $x = 0$, $y = 0$
    for each data predecessor p of j **do**
      if there is a path from $n \rightsquigarrow p$ **then**
        increase a by 1
      if there is not a path $s \rightsquigarrow p$ **then**
        increase x by 1
      increase c by 1
    for each data successor i of j **do**
      if there is a path $n \rightsquigarrow i$ **then**
        decrease a by 1
      decrease c by 1
    if $x \neq 0$ **then**
      for each $k \in A[j]$ **do**
        for each data successor m of k **do**
          if $n \rightsquigarrow m$ but not $s \rightsquigarrow m$ **then**
            increase y by 1
    decrease b by $x \cdot y$
  set the priority vector of s under n to (a, b, c)

Figure 10.9: Successor Priority Assignment

procedure ScheduleDFS(n)
  if n has not been visited **then**
    add n to the visited set
    for each ctrl. succ. i of n in descending priority **do**
      ScheduleDFS(i)
    for each data successor i of n **do**
      ScheduleDFS(i)
    insert n at the beginning of the schedule

Figure 10.10: The Scheduling Procedure

1: **procedure** Restructure
2:   Clear the currently-active branch of each fork
3:   Clear master-copy(n) and latest-copy(n) for each node n
4:   **for each** n in scheduled order starting at the root **do**
5:     $D = $ DuplicationSet(n)
6:     **for each** node d in D **do**
7:       DuplicateNode(d)
8:     **for each** node d in D **do**
9:       ConnectPredecessors(d)

Figure 10.11: The Restructure procedure

out. The path n2 → n3 → n4 → n7 affects the second priority number of the left branch. Under our definitions, the right branch has highest priority and will be visited first during the depth-first search used for scheduling.

Similarly, node n9 will be visited before n7 because the first priority number of n7 is smaller due to the data arc n10 → n11. Finally, n5 will be visited after n4 because n5 has minimum priority.

Constructing the schedule

The scheduling algorithm (Figure 10.10) uses a depth-first search to topologically sort the nodes in the PDG. The control successors of each node are visited in order from highest to lowest priority (assigned by Figure 10.9). Ties are broken arbitrarily, and data successors are visited in an arbitrary order. The label on each node in Figure 10.7 indicates its position in the schedule: n1 is first, followed by n2, n3.

10.2.3 Restructuring the PDG

The scheduling algorithm presented in the previous section totally orders all the nodes in the PDG. Data dependencies often force the execution of subgraphs under fork nodes to be interleaved (control dependencies cannot directly induce interleaving because of the PLCA rule). The algorithm

```
1:  function DuplicationSet(n)
2:      D = {n}
3:      Clear the visited set
4:      DuplicationVisit(n)
5:      return D

6:  function DuplicationVisit(n)
7:      if n has not been visited then
8:          Mark n as visited
9:          if latest-copy(n) is undefined then
10:             Include n in D
11:         for each predecessor p of n do
12:             if p is a fork and p → n is not currently active then
13:                 Include n in D
14:             if DuplicationVisit(p) then
15:                 Include n in D
16:     return true if n ∈ D
```

Figure 10.12: The DuplicationSet function. A node is in the duplication set if it is along a path from a fork node that leads to n but whose active branch does not.

described in this section restructures the PDG by inserting guard variables (specifically, assignments to and tests of guard variables) according to the schedule to produce a PDG where the subgraphs under fork nodes are never interleaved.

The restructuring algorithm does two things: it identifies when a subgraph must be cut away from an existing subgraph according to the schedule and reattaches the cut subgraphs to nodes that test guard variables to ensure the behavior of the PDG is preserved.

The restructure procedure

The Restructure procedure (Figure 10.11) steps through the nodes in scheduled order, adding a minimal number of nodes to the graph under construction that ensures each node in the schedule can be executed without interleaving the execution of subgraphs under any fork. It does this in three phases for each node. First, it calls DuplicationSet (Figure 10.12, called from line 5 in Figure 10.11) to establish which nodes must be duplicated in order to reconstruct the control flow to the node n. The boundary between the set D and the existing graph can be thought of as a cut. Second, it calls DuplicateNode (Figure 10.13, called from line 7 of Figure 10.11) on each of these nodes to create new predicate nodes that reconstruct control using a previously-cached result of the predicate test. Finally, it calls ConnectPredecessors (Figure 10.14, called from line 9 of Figure 10.11) to connect the

1: **procedure** DuplicateNode(n)
2:   **if** n is a fork or a statement **then**
3:     Create a new copy n' of n
4:   **else** {n is a predicate}
5:     **if** master-copy(n) is undefined **then** {making first copy}
6:       Create a new copy n' of n
7:       master-copy(n) = n'
8:     **else** {making second or later copy}
9:       Create a new node n' that tests v_n
10:      **if** master-copy(n) = latest-copy(n) **then** {second copy}
11:        **for** $i = 0$ to (the number of successors of n) $- 1$ **do**
12:          Create a new statement node a' assigning $v_n = i$
13:          Attach a' to the ith successor of master-copy(n)
14:        **for each** successor f' of master-copy(n) **do**
15:          Find a', the assignment to v_n under f'
16:          Add a data-dependence arc from a' to n'
17:      Attach a new fork node under each successor of n'
18:   **for each** successor s of n **do**
19:     **if** s is not in D **then**
20:       Set latest-copy(s) to undefined
21:   latest-copy(n) = n'

Figure 10.13: The DuplicateNode procedure. This makes either an exact copy of a node or tests cached control-flow information to create a node matching n.

1: **procedure** ConnectPredecessors(n)
2:   Let n' = latest-copy(n)
3:   **for each** predecessor p of n **do**
4:     Let p' = latest-copy(p)
5:     **if** p is a fork **then**
6:       Add a new successor $p' \to n'$
7:       Mark $p \to n$ as the active branch of p o
8:     **else** {p is a predicate}
9:       **for each** arc of the form $p \to n$ **do**
10:        Let f' be the corresponding fork under p'
11:        Add a successor $f' \to n'$

Figure 10.14: The ConnectPredecessors procedure. This connects every predecessor of n appropriately, possibly using nodes that were just duplicated. As a side-effect, it remembers the active branch of each fork.

predecessors of each of the nodes in the duplication set, which incidentally includes n, the node being synthesized.

The main loop in Restructure (lines 4–9) maintains two invariants. First, each fork maintains its currently-active branch, i.e., the successor in whose subgraph a node was most recently added. This information, tested in line 12 of Figure 10.12 and modified in line 7 of Figure 10.14, is used to determine whether a node can be added to an existing part of the new graph or whether the paths leading to it must be partially reconstructed to avoid introducing interleaving.

The second invariant is that the latest-copy array holds, for each node that appears earlier in the schedule, the most recent copy of each node. The node n can use these latest-copy nodes if they do not come from forks whose active branch does not lead to n.

The DuplicationSet function

The DuplicationSet function (Figure 10.12) determines the subgraph of nodes whose control flow must be reconstructed to execute the node n. It is a depth-first search that starts at the node n and works backward to the root. Since the PDG is rooted, all nodes in the PDG have a path to the root node and therefore DuplicationVisit traverses all nodes that are along any path from the root to n.

A node n becomes part of the duplication set D under three circumstances. The first case, tested in line 9, occurs when the latest copy of a node is undefined, which can happen when a node is duplicated but its successor is not. lines 18–20 (Figure 10.13) clear the latest-copy array for the successors of a node.

The second case, tested in line 12, is when the immediate predecessor p of n is a fork but n is not the currently active branch of the fork. This indicates that to execute n would require interleaving because the PLCA rule tells us that there cannot be a path to n from p through the currently-active branch under p.

The final case, line 14, occurs when any of n's predecessors are also in the duplication set.

As a result, every node in the duplication set D is along some path that leads from a fork node f to n that goes through a non-active branch of f, or leads from a node that has not been copied "recently." These are exactly the nodes that must be duplicated to reconstruct all paths to n.

The DuplicateNode procedure

Once the DuplicationSet function has determined which nodes must be duplicated to reconstruct the control paths to node n, the DuplicateNode procedure (Figure 10.13) actually makes the copies. Duplicating statement or fork nodes is trivial (line 3): the node is copied directly and the latest-copy array is updated (line 21) to reflect the fact that this new copy is the most recent version of n, something that is later used in ConnectPredecessors. Note that statement nodes are only ever duplicated once, when they appear in the schedule. Fork nodes may be duplicated multiple times.

The main complexity in DuplicateNode comes when n is a predicate (lines 5–17). The first time a predicate is duplicated (i.e., the first time it appears in the schedule), the master-copy array entry for it is undefined (it was cleared at the beginning of Restructure—line 3 of Figure 10.11), the node is copied directly, and this copy is recorded in the master-copy array (lines 6–7).

After the first time a predicate is duplicated, its duplicate is actually a predicate node that tests v_n, a variable that stores the decision made at the predicate n (line 9). There is just one special case: the second time a predicate is copied (and only the second time—CEC does not want to add these assignments more than once), assignment nodes are added under the first copy (i.e., the master-copy of n in the new graph) that save the result of the predicate in the v_n variable. This is done in lines 11–13.

An invariant of the DuplicateNode procedure is that every time a predicate node is duplicated, the duplicate version of it has a new fork node placed under each of its successors (line 17). While these are often redundant and can be removed, they are useful as an anchor point for the nodes that cache the results of the predicate and in the uncommon (but not impossible) case that the successor of a predicate is part of the duplicate set but that the predicate is not.

The ConnectPredecessors procedure

Once DuplicateNode runs, all nodes needed to run n are in place but unconnected. The ConnectPredecessors procedure (Figure 10.14) connects these duplicated nodes to the appropriate nodes.

For each node n, ConnectPredecessors adds arcs from its predecessors, i.e., the most recent copies of each. The only minor trick occurs when the predecessor is a predicate (lines 9–11). First, DuplicateNode guarantees (line 17 of Figure 10.13) that every successor of a predicate is a fork node, so ConnectPredecessors actually connects the node to this fork, not the predicate itself. Second, it can occur that a single node can have a particular predicate node appear two or more times among its predecessors. The *foreach* loop in lines 9–11 connects all of these explicitly.

Examples

Running this procedure on Figure 10.7 produces the graph in Figure 10.15. The procedure copies nodes n1–n5. At this point, n0 → n3 is the active branch under n0, which is not on the path to n6, so a cut is necessary. DuplicationSet returns {n1, n6}, so n1 will be duplicated. This causes

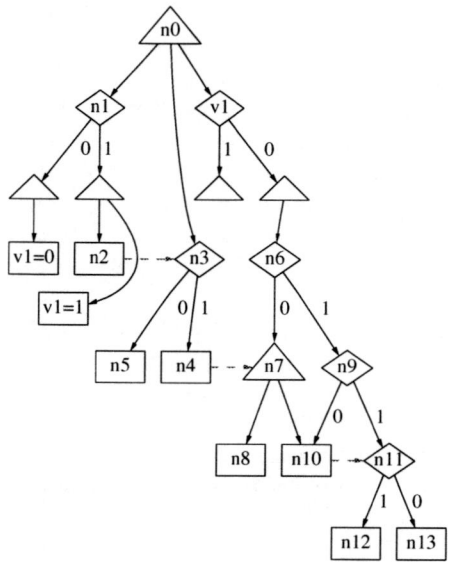

Figure 10.15: The restructured PDG from Figure 10.7. This example only adds the single guard variable v1. Some unary fork nodes generated by Restructure have been omitted for clarity.

DuplicateNode to create the two assignments to v1 under n1 and the test of v1. ConnectPredecessors then connects the new test of v1 to n0 and n6 to the test of v1. Finally, the algorithm just copies nodes n7–n13 into the new graph.

Figure 10.16 illustrates the operation of the procedure on a more complicated example. The PDG in (a) has some bizarre control dependencies that force the nodes to be executed in the order shown. The dizzying number of forced interleavings generates a fairly complex final result, shown in Figure 10.16e.

The algorithm behaves simply for nodes n0–n8. The state after n8 has been added is shown in (b).

Adding n9, however, is challenging. DuplicationSet returns {n9, n6, n5} because n8 is the active node under n4, so DuplicateNode copies n9, makes a second copy of n6 (labeled n6'), creates a new test of v5, and adds the assignments to v5 under n5 (the fork under the "0" branch from n5 has been omitted for clarity). Adding n9's predecessors is easy: it is just the new copy of n6, but adding n6's predecessors is more complicated. In the original graph, n6 is connected to n3 and n5, but only n5 was duplicated, so n6' is connected to v5 and to a fork off the copy of n3.

Figure 10.16d adds n10, which is simple because although n3 was the

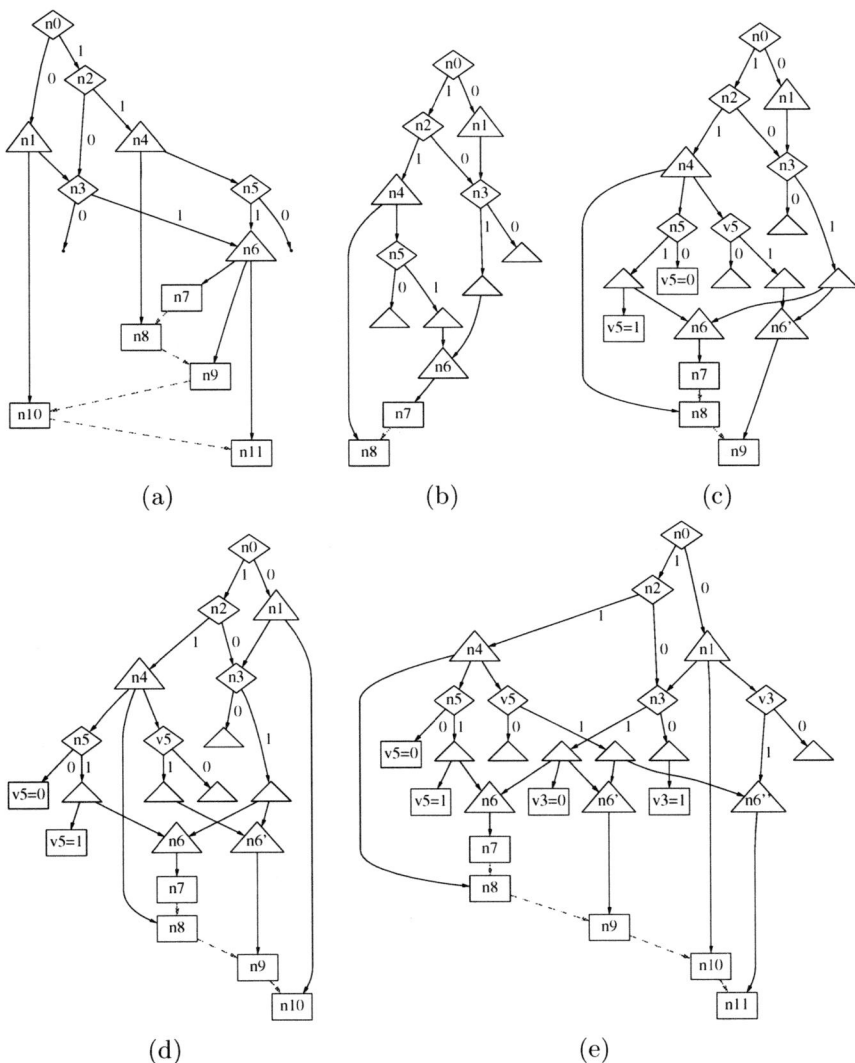

Figure 10.16: (a) A complex example. (b) After adding nodes n0–n8. (c) After adding n9, (d) n10, and (e) n11.

active branch under n1, n10 only has it as a predecessor.

Finally, (e) shows the addition of n11, completing the graph. DuplicationSet returns {n11, n6, n3}, so n3 is duplicated and assignment nodes to v3 are added. Again, n6 is duplicated to become n6″, but this time n3 was duplicated.

Fusing guard variables

An unfortunate choice of schedule clearly illustrates the need for guard variable fusion. Consider the correct but non-optimal schedule n0, n1, n2, n6, n9, n3, n4, n5, n7, n8, n10, n11, n12, n13 for the PDG in Figure 10.7. Figure 10.17 depicts the effect of so many cuts. The main waste is the cascade of conditionals along the right side of the graph (predicates on v1, v6, and v9). For efficiency, we replace such predicate cascades with single multi-way conditionals.

Figure 10.18 illustrates the effect of fusing guard variables. The predicate cascade has been replaced by a single multi-way branch that tests the fused guard variable v169 (formed by fusing predicates v1, v6, and v9). Similarly, group assignments to these variables are fused, resulting in three single assignments to v169 instead of three group concurrent assignments to v1, v6, and v9.

10.2.4 Generating Sequential Code

After the restructuring procedure described above, the structure of the PDG allows the subgraphs under each fork node to be executed in a particular order. This order is non-obvious when there is reconvergence in the graph, and appears to be costly to compute. Fortunately, Simons and Ferrante [67] developed the external edge condition (EEC) as an efficient way to compute this ordering. Basically, the nodes in $eec(n)$ are executed whenever any node in the subgraph under n is executed.

In what follows, $X < Y$ denotes $G(X)$ must be scheduled before $G(Y)$; $X > Y$ denotes $G(X)$ must be scheduled after $G(Y)$; $Y \sim X$ denotes any order is acceptable; $Y \neq X$ denotes no order is acceptable. Here, $G(n)$ represents n and all its control descendants.

CEC reconstructs the graph by ordering fork successors. Given the EEC information, it uses the rules in Steensgaard's decision table [68] to order pairs of fork successors. When the table says any order is acceptable, we order the successors based on data dependencies. However, if, say, the EEC table says $G(X)$ must be scheduled before $G(Y)$, yet the data dependencies indicates the opposite order, the data dependencies win and two additional nodes are inserted, one that sets a guard variable and the other that tests it. Figure 10.19 illustrates the procedure.

In Figure 10.15, data dependency forces n11 > n10, but the external edge condition could require n10 > n11 if there were a control edge from a

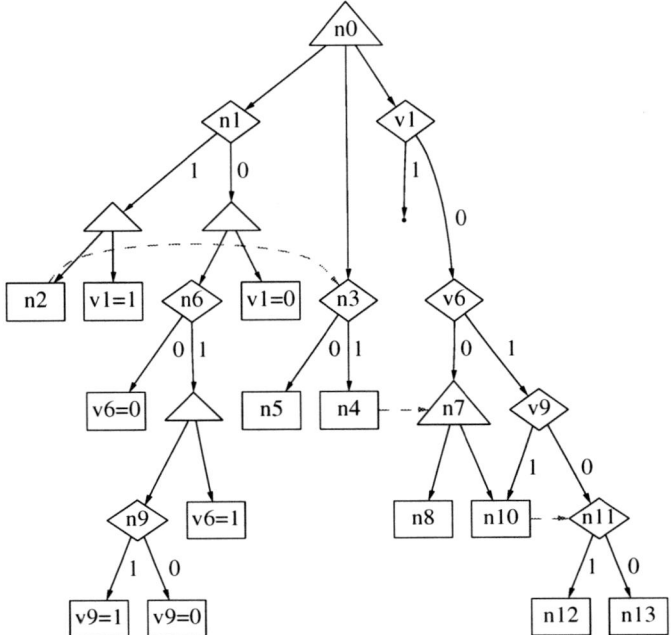

Figure 10.17: The reconstructed PDG from Figure 10.7 induced by a different schedule.

descendant of n11 to a descendant of n10 (i.e., if there were more nodes under n10). In this case, n10 ≠ n11, so our algorithm will cut the graph at n11 and add a guard there.

This produces a sequential control-flow graph for the concurrent program. CEC generates structured C code from it using the algorithm described in Edwards [27].

10.3 Benchmarks

This section presents some experimental results for the two compilation schemes presented in this chapter.

The dynamic approach

Table 10.1 shows the experimental speed results for the dynamic technique implemented in the CEC compiler. Table 10.2 provides some statistics on the Esterel examples.

The results are mixed: Potop-Butucaru's grc2c beats CEC on four of the five examples, but CEC is substantially faster on the largest example,

Example	CEC1	CEC2	grc2c	Saxo	(fast)	EC	V3	V5
1-tcint	0.28	0.34	**0.14**	0.34	0.25	0.18	0.25	1.3
2-wristwatch	0.78	0.93	**0.61**	0.89	0.87	0.86	0.62	2.1
3-atds100	0.11	0.13	**0.03**	0.11	0.08	0.10	-	66.0
4-mca200	1.66	2.75	**1.47**	2.62	2.35	1.79	-	29.0
5-chorus	**0.94**	1.52	1.54	1.42	1.29	1.76	-	51.0

Table 10.1: Experimental results for the dynamic approach. Time, in seconds, to run 1 000 000 iterations of the generated code on a 1.7 GHz Pentium 4 (shorter is better, best values are emphasized). The columns CEC1 and CEC2 respectively correspond to the CEC compiler generating computed *goto*s or *switch* statements. The (fast) column is for the fast version of the Saxo-RT compiler. EC is the Synopsys compiler of Edwards. V3 represents FSM code generated by the INRIA compiler, which can only be generated for two of the examples. The V5 column lists times for code generated by the circuit code generated by the INRIA compiler.

Example	Size	Clusters	Levels	C/L	Threads
1-tcint	357	101	19	5.3	85
2-wristwatch	360	87	13	6.7	87
3-atds	622	156	16	9.8	138
4-mca200	5354	148	15	9.9	135
5-chorus	3893	662	22	30.1	563

Table 10.2: Statistics for the examples. Size is the number of Esterel source lines after *run* statement expansion and pretty-printing. Clusters is the number of clusters found by the algorithm in Figure 10.6. Levels is the number of levels the clusters were compressed into. C/L is the ratio of clusters to levels. Threads is the number of concurrent threads as reported by the Synopsys compiler [27].

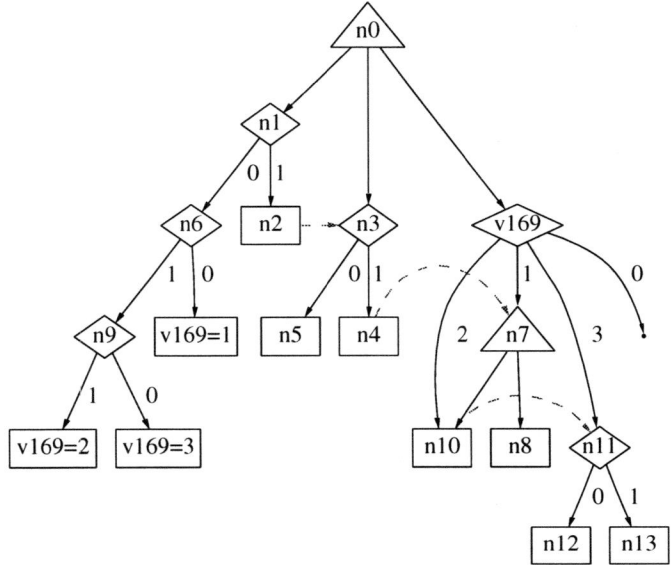

Figure 10.18: The PDG of Figure 10.17 after guard variable fusion

Chorus. Furthermore, CEC is faster than the Saxo-RT compiler on the three largest examples (5-chorus, 4-mca200, and 2-wristwatch). This is expected: the CEC technique should become faster than the Saxo-RT compiler on larger examples since our (similar) technique has less overhead for unexecuted parts of the program.

The number of clusters and levels in Table 10.2 suggests why the CEC technique is better on some programs and worse on others. A key trick is the use of one linked list per level for scheduling. The more clusters there are per level (measured, e.g., by the C/L average in Table 10.2), the more the dynamic technique differentiates itself from the Saxo-RT compiler. The results bear this out: 5-chorus, which has the largest number of clusters per level on average, exhibits the largest improvement over the other techniques.

The circuit code generated by the INRIA compiler for these examples produces far worse run-times. The results for 2-wristwatch show the least variance because it calls the most user-defined functions, something none of these compilers attempt to optimize.

These timing results were obtained by applying a random sequence of inputs to the code generated by each compiler and measuring the time it took to execute 1 000 000 reactions. Note that the ratio of these measured times differ noticeably from those reported in Section 9.4. This can be attributed to a variety of factors including a different processor (Potop-Butucaru used a P3, these results are on a P4), different C compilers and optimization flags,

procedure OrderSuccessors(G)
   for each node n **do**
     if n is a fork node **then**
       original-successors = control successors of n
       clear the control successors of n
       for each X in original-successors **do**
         for each control successor Y of n **do**
           if $X \sim Y$ **then**
             if $\exists (m,n) \in D,\ m \in G(X), n \in G(Y)$ **then**
               insert X before Y in n's successors
           else if $Y < X$ **then**
             if $\exists (m,n) \in D,\ m \in G(Y), n \in G(X)$ **then**
               Cut Y
             insert X before Y in n's successors
           else if $Y > X$ **then**
             if $\exists (m,n) \in D,\ m \in G(X), n \in G(Y)$ **then**
               Cut X
             else
               insert X before Y in n's successors
           else if $Y \neq X$ **then**
             if $\exists (m,n) \in D,\ m \in G(X), n \in G(Y)$ **then**
               Cut Y
             insert X before Y in n's successors
             else
               Cut X
         if X was not inserted **then**
           append X to the end of n's successors

Figure 10.19: The successor ordering procedure

Example	Lines	Average cycle time (seconds)		
		V5	CEC1	CEC3
1-tcint	687	11	2.8	**2.4**
3-atds100	948	45	7.7	**1.3**
multi6	113	10	2.3	**1.4**
multi8	62	1.1	1.7	**0.63**
greycounter	82	6.0	3.9	**0.94**
abcd	111	5.2	**1.5**	1.7

Table 10.3: Experimental Results. CEC3 is the PDG-based approach. CEC1 is the dynamic technique implemented in CEC.

and perhaps different stimulus.

The PDG-based approach

Here are some experiments that compare the speed of the code generated by this technique to that of the other CEC compilation line, and to the reference circuit code generated by the INRIA compiler.

To obtain the average cycle times shown in Table 10.1, we ran the generated C code from all three compilers (compiled with gcc -O3) for 10 million cycles on a 2.5 GHz Intel Pentium 4 running Linux. Most examples are fairly small, but 1-tcint and 3-atds100 (both bus controllers) are reasonably large and, we believe, illustrative of our technique.

A

Language Extensions

The full Esterel V5 language includes the Kernel Esterel primitives and derived statements whose semantics are given by structural expansion into the primitives. The full language also includes a number of constructs that extend the primitives in non-elementary ways. We describe these here along with semantic and code generation hints.

A.1 Signal Expressions

In Kernel Esterel, presence or preemption tests only involve single signals. In the full language, such tests may involve arbitrarily complex signal expressions constructed using the operators **and**, **or**, **not**, and **pre**, parenthesis (), and the predefined signal **tick**, which is always present. Some examples:

- Await the next instant where A and B are both present

    await [A and B]

- If C is present and none of A, B, or C were present in the previous instant, then start p, else start q

    present [C and not pre(A or B or C)] then p else q end

- Abort the execution of p the instant following the one where A is first present

    abort p when pre(A)

- Await the next clock tick. Equivalent to **pause**.

    await tick

A.1.1 Syntactic Aspects and Limitations

Esterel uses the usual precedence rules to limit the number of parentheses necessary in signal expressions. The **not** operator binds most tightly, followed by **and** and finally **or**. So the following two expressions are equivalent.

```
not A and B or C            ((not A) and B) or C
```

The **pre** operator is not subject to precedence rules; its argument must always be enclosed in parenthesis.

There are many limitations on the use of **pre**: only the INRIA compiler can handle it, and that compiler does not allow nested **pre** operators, so, e.g., `pre(pre(A))` is not allowed; **pre** cannot be used on sensors, task return signals, or traps; and `pre(tick)` is not allowed.

A signal expression must be enclosed in brackets except for two cases: when it is the argument of a **present** statement and when it consists of a single **pre** operator.

A.1.2 Combinational Expressions

We start the formal description of signal expressions by considering expressions that only include the **and**, **or**, and **not** operators.

Implementation

The simplest way of giving the formal meaning of combinational expressions is by translation into digital circuits. The translation of a signal expression test fits perfectly into the framework defined in Chapter 6. The only difference is that the *Test* sub-circuits represent here the complex test expression instead of a simple signal or data test. We illustrate the translation process with a simple example that shows how the *Test* sub-circuit is obtained from the signal expression through a simple mapping. Consider the statement

```
present [C and not (A or B or C)] then p else q end
```

The corresponding *Test* sub-circuit is pictured in Figure A.1. The test gates (gray) are driven by the dashed box, which contains the sub-circuit corresponding to the signal expression. The structural correspondence between Esterel operators and gates is obvious. Note that the *Test* subcircuit has an input for each signal it uses.

Things are even simpler for GRC, as test expressions are simply copied inside the corresponding **Test** node. The structural translation rules of the statements **present** and **suspend**, given in Section 8.2, are amended by changing the definition of the signal link sets. In both cases, instead of adding a single signal input port to the corresponding signal link set, we need to add one port per signal used in the signal expression and not covered by a **pre**.

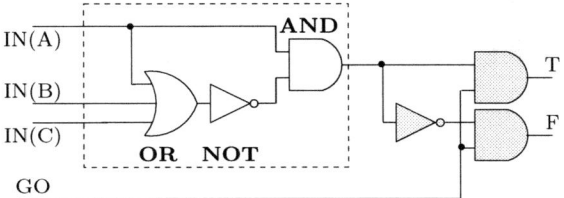

Figure A.1: Test sub-circuit for the expression "C and not (A or B or C)"

Semantics by kernel expansion

To remain inside the Esterel semantic framework, we also give meaning to combinational expressions by translation into Kernel Esterel. While not useful in practice, this transformation illustrates how a digital circuit can be encoded in Esterel.

The first step is to reduce the translation problem to the case of simple present statements. We do this by introducing an auxiliary signal, for example, for an abort statement:

    abort p when *signal-expr*

which becomes

    trap T in signal AUX in
      abort p when AUX ; exit T
    ||
      loop present *signal-expr* then emit AUX end ; pause end
    end end

At this point, our problem is to encode into Kernel Esterel statements the instantaneous computation of $\text{AUX} = \text{signal-expr}(S_1, \ldots, S_k)$, where S_i, $1 \leq i \leq k$ are the signals involved in *signal-expr*. A brute force encoding suffices: associate one new signal to each sub-expression of *signal-expr* that is not an existing signal. Let $S_{k+1}, \ldots, S_n$ be the new signals, and assume that S_n corresponds to the global expression *signal-expr*. Then put in parallel the computations of all sub-expressions S_i, $k+1 \leq i \leq n$ from their immediate factors:

    signal $S_{k+1}, \ldots, S_n$ in
      present S_n then emit AUX end
    ||
      the factors, put in parallel
    end

where

- if $S_i = S_j$ or S_k, the factor for S_i is

    present S_j then emit S_i end
    ||
    present S_k then emit S_i end

- if $S_i = S_j$ and S_k, the factor for S_i is

    present S_j then present S_k then emit S_i end end

- if $S_i = $ not S_j, the factor for S_i is:

    present S_j else emit S_i end

Consider *signal-expr*=A and not B. In this case, the statement

    present A and not B then emit AUX end

can be replaced with its kernel expansion

    signal S_1, S_2 in
      present S_2 then emit AUX end
    || % the factor for S_2 = A and S_1
      present A then present S_1 then emit S_2 end end
    || % the factor for S_1 = not B
      present B else emit S_1 end
    end

A.1.3 The pre Operator

pre(S) gives the present/absent status of the signal S in the previous instant. Note that here, "previous execution instant" refers to the base clock of the statement in whose scope S is defined, not the global clock. The previous status pre(S) can be seen as signal with the same scope as S: a local declaration or the entire module. The formal semantics of pre is given by kernel expansion:

    signal S in p end

becomes

    trap T in
      signal S, preS in
        p' ; exit T
      ||
        loop
          present S then
            pause ; emit preS

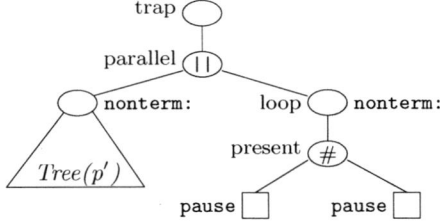

Figure A.2: The tagged selection tree generated for the kernel expansion of pre .

```
      else
        pause
      end present
    end loop
  end signal
end trap
```

where p' is obtained from p by replacing all occurrences of pre(S) by preS.

Translating the pre operator using this expansion rule results in redundant code. Consider the case of the GRC code whose selection tree is given in Figure A.2. In the light of the analysis realized in Section 8.4, the associated GRC flowgraph can be simplified and C code can be generated (cf. Section 9.2) where the encoding of the two pause statements is performed on a single bit of the state encoding. This corresponds to the intuition that the Boolean status of a signal is preserved from an instant to the next using a single state element. Similar transformations can be performed at netlist level, where only one register suffices to preserve the signal status.

Valued signals

The pre operator can also return the previous value of a valued signal: if S is a valued signal, pre(?S) is the value of S in the previous execution instant. Again, the formal semantics is given by kernel expansion:

```
signal S : TYPE in p end
```

becomes

```
trap T in
  signal S, preS : TYPE in
    p' ; exit T ||
    loop
      present S then
        var v := ?S : TYPE in pause ; emit preS(v) end
```

```
            else
              pause
            end present
          end loop
        end signal
      end trap
```

where p' is obtained from p by replacing all occurrences of `pre(S)` by `preS`, and all occurrences of `pre(?S)` by `?preS`.

The value `pre(?S)` should not be used before the initialization of `?S` (through declaration or first emission). However, current implementations of Esterel V5 do not check this. The return value of such illegal reads is unspecified. The Esterel V7 compiler of Esterel Studio does check the initialization and can formally verify that a non-initialized signal value is never read.

Signal expressions and `pre`

The operator `pre` can be applied to any combinational expression. To define the semantics of `pre` on the simple negation `pre(not S)`, we develop on the expansion of `pre(S)`, given above. We introduce another signal `preNotS` with the same scope as `S` and `preS`. Then, we replace the loop defining `preS` with

```
      loop
        present S then pause ; emit preS
        else pause ; emit preNotS
        end present
      end loop
```

Finally, the definition of p' also changes: all occurrences of `pre(not S)` are replaced by `preNotS`.

When the argument of `pre` is a more complex combinational expression, we start by using De Morgan's laws to move `not` operators to the leaves (signals). For instance, `not (B or not C)` becomes `not B and C`. Then, the following two expansion rules move the `pre` operators onto the leaves:

$\text{pre}(expr_1 \text{ and } expr_2) = \text{pre}(expr_1) \text{ and } \text{pre}(expr_2)$

$\text{pre}(expr_1 \text{ or } expr_2) = \text{pre}(expr_1) \text{ or } \text{pre}(expr_2)$

This completes the definition of the semantics of `pre`.
pre@pre!dangers of

Dangers

The `pre` operator should be used with caution. First of all, adding a `pre` adds not only a register, but also the complex logic controlling it, as the following example shows. Here, we emit the signal `A`, and in the next execution instant we test `pre(A)`.

```
    emit A ; pause ; present pre(A) then emit O end
```

While intuition may tell us that O is emitted in the second instant regardless of the context, this is not true. Indeed, if we suspend the statement for one instant (with `emit S`), then the test on `pre(A)` will not emit O.

```
signal A,S in
  pause ; emit S
||
  suspend
    emit A ; pause ; present pre(A) then emit O end
  when S
end
```

Maybe more important, if a signal AUX is defined in the scope of p

```
signal AUX in
  loop
    present expr then emit AUX ; pause else pause end
  end
||
  p
end signal
```

the expressions `pre(AUX)` and `pre(`*expr*`)` may not be equivalent in p, for example, if `suspend` statements separate the declaration of AUX from the declarations of the signals in *expr*. In other words, `pre` over complex expressions cannot be handled with traditional retiming.

A.1.4 Delay Expressions. Preemption Triggers

Delay expressions are the triggers of the preemption statements such as `suspend` or `abort`. A delay expression is a signal expression, possibly prefixed with one modifier. In this section, we present the syntax and intuitive meaning of delay expressions. Their exact semantics depends on the statement that uses them; details are in Appendix B.

When the delay expression is a simple signal expression, the preemption test is performed from the *second* execution instant on.

```
    await signal-expr
```

For instance, if *signal-expr* is true in the start execution instant, the given statement will discard it.

When the signal expression is prefixed with `immediate`, the test is also performed in the first instant.

```
    await immediate signal-expr
```

If *signal-expr* is true in the start execution instant, the statement will instantly terminate.

When the signal expression is prefixed with an integer data expression, we say that we have a counter (or counted) delay. Counter delays facilitate the programming of behaviors where preemption occurs after a given number of occurrences of the same event. The `immediate` modifier cannot be used with a counted delay. For example,

    await c+3 [A and B]

When this statement is started, the integer expression c+3 is evaluated. If it not positive, it is taken to be 1. This value is used to initialize a counter that is decremented on every occurrence of the given event (here, both A and B present in the same instant) that occurs after the first instant. When the counter reaches zero, the `await` terminates.

A.2   Traps and Trap Expressions

Valued traps are like valued signals. Defining the semantics and code generation for the complex variants of the `trap` and `exit` statements requires a non-trivial expansion.

The simplest form of the `trap` statement is the primitive form defined earlier

    trap T in p end

This simple form defines an alternate exit point, named T, from statement p. When

    exit T

is executed inside p, the entire `trap` construct terminates and passes control to the next instruction in sequence.

The trap declaration statement also allows the definition of a trap handler statement q, executed if the trap is exited, but not when the body p terminates normally or is preempted.

    trap T in p handle T do q end

A kernel expansion of this statement requires a non-trivial rewriting of both the `trap` construct and any "exit T" statements in q. We need to introduce a new trap for cases where the trap body terminates normally (code 0). Assume that trap name U is available for this trap. In this case, the trap construct is expanded into

    trap U in
      trap T in
        p ; exit U

```
    end ;
  q
end
```

A.2.1 Concurrent Traps and Trap Expressions

Nested traps have an implicit precedence. When two traps are exited simultaneously, the outermost has higher priority and hides any inner ones. The full Esterel language also provides so-called concurrent traps to allow two or more to be visible at once. Such traps are declared using a single **trap** statement.

```
trap T₁,...,Tₙ in
  p
handle trap-expr₁ do q₁
...
handle trap-exprₖ do qₖ
end
```

Here, we declared the concurrent traps $T_1, \ldots, T_n$. All trap names in a concurrent trap definition must be unique.

Of course, defining concurrent exit points is only useful when we can define multiple, concurrent exit points (handlers). Esterel allows users to define handlers that are activated upon Boolean combinations of traps. More precisely, handler triggers are trap expressions obtained by combining the trap identifiers T_i, $1 \leq i \leq n$, with the Boolean connectors **and**, **or**, and **not**. For instance, to activate handler q_1 when both traps T_1 and T_2 are exited inside p, we set $trap\text{-}expr_1$ to be T_1 **and** T_2.

When at least one of the concurrent traps is exited inside the concurrent trap definition, the body p is preempted, all the trap expressions are evaluated, and control is given to all the branches whose handler is true. These branches run in parallel. When all activated handlers have terminated, the **trap** statement terminates and control is given in sequence.

No restriction is imposed on the number of trap handlers or the complexity of the trap handler expressions, but certain compilers warn when a given condition/trap is handled twice.

Semantics by expansion

Unlike normal nested traps, concurrently-defined traps share a single trap code, so no priority is defined among them. Instead, the expansion adds auxiliary signals that indicate which traps were actually raised. In the definition of the following expansion, we assume that the signal names $T_1, \ldots, T_n$, and the trap name T are not already used inside p.

```
signal T₁,...,Tₙ in
  trap T in
```

```
      p'
    end ;
    [
      present trap-expr₁ then q₁ end
    ||
       ...
    ||
      present trap-exprₖ then qₖ end
    ]
  end
```

The unique trap code is represented here by T, while the trap names are represented by the new signals. The body p' is obtained from p by rewriting every instance of "exit T_i" into

```
    emit Tᵢ ; exit T
```

A.2.2 Valued Traps

Like exceptions in other programming languages, Esterel allows traps to carry a data value. More precisely, a type can be specified for every declared trap. The rules for declaring valued traps are the same as those for declaring valued signals (i.e., declared traps can be of simple or combined type, initialized or not). Just like valued signal emission statements, the associated exit statements specify both the trap that is exited and a data value or expression. The value of a trap T can be used only in the handlers of the trap statement that declared T.

For example,

```
    trap T1 : integer, T2 := 3.0 : combine double with +,
         T3 (integer), T4 in
       await I1 do
         exit T1(10)
       ||
         present I2 then exit T2(7.0) end
       end
    ||
       await I3 do exit T2(10.0) end
    ||
       await I4 do exit T3(10) end
    ||
       await I5 do exit T4 end
    handle T1 and     T3 do emit O1(??T1+??T3)
    handle T2             do emit O2(2.0*??T2)
    handle T4 and not T2 do emit O2(??T2)
    handle            not T1 do emit O1(??T1)
    end
```

Three of the four traps declared here are valued. The value of trap T is used in handlers as "??T." As for valued signals, one must take care not to read uninitialized traps that have not been exited since no checks are performed at compile time. For instance, the handler for "not T1" emits an unspecified (random) value.* In the handler for "T4 and not T2," the trap T2 is not exited, but its value is initialized with value 3.0. Finally, note that the type of T2 is combined. Like for valued signals, this means that the values produced by concurrent exit statements are combined in a single trap value, using the provided combine operator "+." When a trap T is not combined, the user must take care not to allow the concurrent execution of several "exit T(...)" statements.

Semantics by expansion

The expansion of the general case valued trap declaration follows from the expansion of the concurrent trap definition. The differences are

- the trap types are copied to the corresponding signals,

- the body p is rewritten into p' by replacing instances of "exit T_i(*data-expr*)" with

    emit T_i(*data-expr*) ;
    exit T ;

- the handler q_i is rewritten into q'_i by replacing instances of "??T_i" with "?T_i."

Again, this assumes that no name conflict appears due to expansion. Otherwise, some renaming must be performed.

A.3 The finalize Statement

The finalize statement was introduced in the experimental Esterel version 6 compiler, and is now part of the proposed Esterel version 7 standard. The finalize statement facilitates the writing of "cleanup code" that needs to be executed when a statement is terminated or preempted. Typical cleanup actions free allocated resources or inform other statements that the finalized statement has terminated. We have included the presentation of finalize in this book because it simplifies the definition of tasks, given in the next section. We did not include it in the reference manual (Appendix B).

---

*As for valued signals, the Esterel V7 compiler checks initialization and makes it possible to formally verify that a non-initialized trap value is never read.

Syntax and intuitive semantics

The `finalize` statement has the following form:

    finalize p with q end

The statements p and q are called the body and the finalizer of the `finalize` construct. Statement q must be instantaneous. When started, the `finalize` statement behaves like p until it terminates or until the entire construct is preempted by some enclosing statement. In this case, the finalizer is executed.

For instance, consider the following fragment, which models a simple bus access protocol. A request is emitted, and the `BusGrant` signal is awaited before starting the communication. The communication may end when `PerformCommunication` terminates, or when the `ExternalInterrupt` signal is set. In both cases, we need to release the bus. The following code guarantees that `BusRelease` is emitted in either case.

    abort
      emit BusRequest ;
      await BusGrant ;
      finalize
        run PerformCommunication
      with
        emit BusRelease
      end
    when ExternalInterrupt

When a complex statement is preempted, the finalizers of the different statements that have control are executed in a bottom-up fashion. The innermost are executed first; the finalizer of the root statement is last.

When an incarnation of a statement terminates and/or is preempted, its finalizer must be executed exactly once, even when several causes may determine its termination, e.g., when several enclosing traps are activated.

Problems

The `finalize` statement induces semantic problems and makes code generation difficult. For both reasons, the framework defined in this book is insufficient to handle the new statement.

The problem is that the `finalize` statement allows control to enter statements that are preempted. From a semantic point of view, this means that a new type of control flow must be added to the COS, which traverses preempted statements before their state is reset to execute the finalizer code. This requires fundamental modifications to both the CBS and the COS.

From a code generation point of view, in both netlists and GRC, the `finalize` statement introduces control paths that do not follow any existing structural rules. Perhaps more importantly, finalizers complicate the causal dependencies of a GRC or netlist specification because they are activated by and give control to many different points depending on their trigger. Finally, it is difficult to generate well-structured efficient software code that enters preempted code.

The solution adopted in the Esterel V7 compiler of Esterel Studio is the systematic duplication of finalize code in all places it can be executed. This reduces the problem to the case treated in this book, where only two control entry points exist for each statement.

A.4 Tasks

A task is a piece of code written in the host language designed to run concurrently with the Esterel program. Tasks are designed to represent complex computations that cannot be represented as functions or procedures because they run for more than a single execution instant. Typical tasks are complex data computations or interactions with the asynchronous environment, such as the ones used in robotics. Tasks are hybrid concepts at the boundary between Esterel's synchronous semantics and the unspecified (possibly asynchronous) environment.

We explained in Chapter 2 that tasks are declared much like procedures. The full syntax is given in Appendix B. A task has two lists of parameters: one passed by reference, the other by value. For example, the **task** keyword declares a task

```
task MyTask(integer, MyType)(integer, string) ;
```

and **exec** statement starts its execution

```
exec MyTask(v1, v2)(10, "abc") return R ;
```

The external routine is launched when the **exec** statement receives control. The Esterel program pauses at the **exec** statement and remains paused until the task completes its execution or is preempted.

If and when the external task completes, the execution shell indicates this instant by raising the return signal R. This synchronization allows the program to react by updating the reference parameter variables with the results of the task computation, so that they are safely recovered; and terminating the **exec** statement and giving control in sequence to the following statement. Note that aborting the **exec** statement must kill the external task to be consistent with Esterel semantics.

Suspending a task sends it a suspension signal in each instant it is suspended. How the task and environment handles this signal is unspecified; it may be difficult for an operating system running the Esterel program and the task asynchronously to correctly interpret sequences of suspension commands. The execution shell may need to convert the instant-wise suspension information into suspend/resume information for the operating system.

248   LANGUAGE EXTENSIONS

The mechanisms for task launching and signaling are largely implementation-dependent. The only constraint is that it must respect the Esterel logical view during the start, completion, and preemption execution instants. The following section presents in an intuitive fashion the synchronization rules ensuring the coherency of the logical view.

A.4.1   Task Synchronization Semantics

The protocol for managing task execution is best understood by representing each of its operations with a separate procedure, which not incidentally mimics the host language task interface.

*Encoding an **exec** statement*

In this section, we consider an Esterel program that contains one task and one **exec** statement:

    exec MyTask(*ref_params*)(*val_params*) ;

Five procedures represent the five task operations:

1. The task is launched by a procedure that takes all the parameters of the task call as by-value parameters. By-value is used exclusively because the task cannot access its reference parameters during the execution of the reaction without interfering with the execution of the Esterel program.

    procedure Launch_MyTask()(*ref_params*, *val_params*);

2. Another procedure updates reference parameters. Called when the task completes, this takes the results of the execution of the task and assigns them to the reference parameters of the task call:

    procedure Update_MyTask(*ref_params*)();

3. Another procedure kills the tasks if it is not already finished and does nothing otherwise.

    procedure Kill_MyTask()();

4. Another procedure suspends the asynchronous execution of the task, assuming it has not already terminated. The execution shell calls this when MyTask is suspended in the Esterel program in the current reaction but was not suspended in the previous reaction.

    procedure Suspend_MyTask()();

TASKS

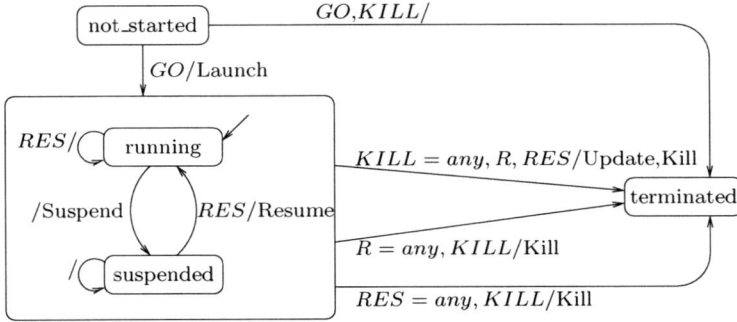

Figure A.3: The life cycle of a task

5. The last procedure resumes the execution of a suspended task, assuming it has not already terminated. The execution shell calls this when the task is resumed after being suspended.

```
procedure Resume_MyTask()();
```

How these procedures are implemented is not specified, only the order in which they are called during the execution of the Esterel program.

With these, we can represent the protocol governing the life cycle of a task in the form of an automaton: Figure A.3. Here, *GO*, *RES*, *KILL* correspond to the Esterel statement being started, resumed, or killed.

The implementation may render suspension and resumption meaningless, e.g., when asynchronous tasks are non-interruptible. In this case, we can use the first three procedures to give an Esterel expansion of the **exec** statement. The expansion uses the **finalize** statement defined in the previous section.

```
finalize
  abort
    call Launch_MyTask()(ref_params,val_params) ;
    halt
  when R do
    call Update_MyTask(ref_params)()
  end
with
  call Kill_MyTask()()
end
```

*Immediate restart of an **exec** statement*

An **exec** statement can be immediately restarted after being aborted or terminated. Consider the fragment

```
loop
  abort
    exec MyTask(...)(...) return R
  when I
end
```

Assume that signal I arrives before the completion of MyTask, or at the same time as R. Then, Esterel calls Kill_MyTask to signal to the environment that the current running instance of MyTask should be killed. Then, it calls Launch_MyTask to signal that a fresh instance of MyTask should be started right away.

In cases where several incarnations of an exec statement are started in a single instant, the rules of reincarnation tell us that at most one of them completes the instant without being killed. If we refer to the protocol of Figure A.3, at most one of the started instances ($GO = 1$) calls Launch and goes to the running state. All the others directly terminate because they are also killed ($KILL = 1$). Therefore, at most two instances of a task can be active in a given reaction. When two instances are active, one is terminating execution and the other is starting.

Tasks and preemption

Using this representation of the task, its is easy to understand its properties. Consider the statement

```
abort
  exec MyTask(ref_params)(val_params) return R
when I
```

If R occurs before I, the reference parameters are updated and the whole statement terminates. If I occurs before R, then the execution of MyTask is aborted. If I and R occur simultaneously, the task completed its execution so that the call to Kill_MyTask will have nothing to do, but Update_MyTask is not called, so the reference variables are not updated with the results. If we replace abort with weak abort in the previous statement, the reference parameters are updated.

Return signal properties

The return signal of an exec statement is bound to the global clock of the program, not to the clock of the exec statement. Therefore, it is possible to test the completion of a task even if it is strongly aborted:

```
abort
  exec MyTask(ref_params)(val_params) return R
when I do
  present R then ... else ... end
end
```

This property of the return signals also leads to less natural situations. If the signals I and R arrive simultaneously in the following fragment, the return signal is lost and the statement remains blocked.

```
suspend
    exec MyTask(ref_params)(val_params) return R
when I
```

In our procedure-based encoding of "exec," this corresponds to the case where the entire construction is suspended when R arrives, so that the control remains blocked on halt.

Return signals can carry data information, as they can be valued signals (even of combined type). There are only two restrictions concerning return signals: no two exec statements of an Esterel program can have the same return signal, and the rule must hold after sub-module instantiation; and return signals cannot be emitted by the Esterel program itself.

Multiple instances of a task

The previously-defined encoding is correct for Esterel programs where each task is executed by a single exec statement. However, when two or more exec statements can launch the same task concurrently, the different instances must be distinguished, using instance identifiers that are given as a parameter to all the synchronization operations (for clarity, we presented the simpler version).

A.4.2 Multiple exec

The multiple exec statement makes it possible to control several tasks simultaneously. It has the following form:

```
exec
case T₁(...)(...) return R₁ do p₁
...
case Tₙ(...)(...) return Rₙ do pₙ
end
```

The section "do p_i" can be omitted if p_i is reduced to nothing.

All tasks are launched simultaneously and concurrently when the statement executes, and it terminates when at least one return signal is received. Any unfinished tasks are aborted. If several return signals have been received, the completed task of lesser index updates its variables, and the exec statement terminates. Due to this exclusiveness property, variables can be shared among reference parameter lists of the n tasks.

B
An Esterel Reference Manual

In this appendix, we define the full Esterel language. Its grammar is given in a variant of the Backus-Naur form; the semantics of Kernel Esterel statements has already been given in the second part of this book. Here, we give the semantics of the derived statements by showing how to expand them into kernel statements.

B.1 Lexical Conventions

B.1.1 Tokens

There are five kinds of tokens: identifiers, keywords, literals, operators, and separators. In general, whitespace (blanks, tabs, newlines, form-feeds) and comments are ignored except that they can separate tokens.

Esterel's scanner uses the usual longest-token rule: if the input stream has been parsed into tokens up to a given character, the next token is the longest string of characters from that point that could be a token.

B.1.2 Comments

Single-line comments start with % and go to the end of a line. If the character immediately following a % is {, this starts a multi-line token that terminates at the first pair of }% characters encountered.

B.1.3 Identifiers

Identifiers are sequences of letters, digits, and underlines (_) and must start with a letter. Identifiers are case-sensitive. No limit is placed on the length of the identifier.

In the definition of the Esterel language grammar, identifiers are covered by the terminal *ident*. As such, it will be used in lists, defined below.

ident-list:
    non-empty-ident-list$_{opt}$

non-empty-ident-list:
    ident
    non-empty-ident-list , ident

More often we use typed identifiers, which are syntactically identical but more suggestive. All named Esterel objects will have such a typed identifier terminal: *signal-ident*, *type-ident*, *function-ident*, *proc-ident* (procedure names), *trap-ident*, *task-ident*, *var-ident*, *constant-ident*, *module-ident*.

B.1.4 Reserved Words

In Esterel, the following words are reserved and cannot be used as identifiers. They include the statement and interface keywords, as well the Boolean literals and predefined functions and procedures. It is important to note that other words, such as `tick` and `integer`, predefined by Esterel in certain namespaces, so their use is subject to restrictions in these namespaces only.

abort	and	await	call	case
combine	constant	copymodule	do	each
else	elsif	emit	end	every
exec	exit	false	function	halt
handle	if	immediate	in	input
inputoutput	loop	mod	module	not
nothing	or	output	pause	positive
present	procedure	relation	repeat	return
run	sensor	signal	suspend	sustain
task	then	timeout	times	trap
true	type	upto	var	watching
weak	when	with		

The `end` keyword, which terminates a statement, can be followed with the statement name. For instance, "`loop p end loop`" is equivalent to "`loop p end`." For `weak abort`, the `end` keyword can be replaced by both "`end abort`" and "`end weak abort`."

B.1.5 Literals

Numeric literals

All numeric literals in Esterel are decimal and unsigned. Their syntax is defined by the following regular expression:

$$[0\text{--}9]^+ \ ([.][0\text{--}9]^*)? \ ([\text{e}|\text{E}][-|+]?[0\text{--}9]^+)?$$

When no decimal dot and no exponent part are present, the literal is interpreted as integer. When a fraction or an exponent part is present, the literal

is floating-point. A floating literal is treated as a float if it is immediately
followed by an f and is otherwise treated as a double.

    integer: 1973, 0
    float: 17e-1f, 0.11f, 0.314e1f
    double: 17e-1, 0.11, 0.314e1

String literals

Strings are single-line sequences of characters delimited by double quotes.

```
"this is a string literal"
"Hello, world!"
```

The only escape sequence is "", which is transformed into a single double quote sign. Example:

```
"""Hello, world!""" is a string literal"
```

Different Esterel compilers may introduce various limitations on string size. Appendix C lists such limitations for the INRIA compiler.

Boolean literals

The two Boolean literals are keywords:

boolean-literal: one of
        false   true

B.2 Namespaces and Predefined Objects

There are eight namespaces in Esterel V5: signals and sensors, traps, variables and constants, functions, procedures, tasks, types, and modules. Thus, the same name can be given to an entity of each of these types without error. Here is a nonsensical, but correct, illustration.

```
module A:

type A;
function A() : integer;
procedure A(integer)(boolean);
task A(integer)(integer);
constant A : integer;
input A;

trap A in
  signal A in            % Hides input A
    var A : integer in % Hides constant A
```

```
            nothing
          end var
        end signal
      end trap

    end module
```

By contrast, the Esterel V7 dialect uses a single namespace so no identifier can be reused for another object in the same scope.

B.2.1 Signals and Sensors

Signals and sensors are Esterel's fundamental objects. There are three main types: pure signals, which only carry presence and absence information; sensors, which only carry a data value; and valued signals, which carry both presence/absence information and a value.

Pure and valued signals may be part of the interface of a module, or may be declared within the module using the `signal` construct. Sensors may only be declared as part of an interface.

The presence/absence status of pure and valued signals does not persist between instants. A signal is absent in an instant unless an *emit* statement (or the equivalent) for the signal runs. However, the data associated with a valued signal does persist between instants, i.e., once it is set, its value persists until set differently.

The value of a sensor may change at the beginning of each instant; it is under the control of the environment.

The value of valued signals and sensors is accessed with the ? operator. The value of a valued signal can be set by the `emit` statement.

All but one signal is user-defined. The predefined signal `tick` is always present in each instant.

signal-name:
> *signal-ident*
> `tick`

B.2.2 Variables and Constants

Variables are objects with a name and a type whose value may be assigned. Constants are named objects whose value is set once when the program starts running, either by the program or by the environment. The value of a constant may be read like a variable but not assigned. Variables are declared within a module; constants are declared as part of the interface of a module.

B.2.3 Traps

A trap is a named exit point for a block of code. There are two types of traps: pure traps, which do no carry a value, and valued traps, which carry a value. The value of a valued trap can be read with the ?? operator and is set by the **exit** statement.

B.2.4 Types

Esterel predefines five basic types: **boolean**, **integer**, **float**, **double**, and **string**. The user can define named types, whose implementation must be provided in the host language at code generation time. To Esterel, all types are abstract, type checking is always strict, and no implicit conversion is performed, even among predefined numerical types. Objects of the same type can be assigned and compared.

type-name:
    type-ident

Predefined type names are not keywords and can be reused. For instance, it is possible, but inadvisable, to call a signal **float** or **integer**.

B.2.5 Functions and Procedures

The user can define functions and procedures, whose definitions are given in the host language. A function has a return type and zero or more pass-by-value arguments. It can be used in expressions. A procedure has zero or more pass-by-value arguments and zero or more pass-by-reference arguments (usually results). Procedure calls are primitive statements starting with the **call** keyword. Functions and procedures are assumed to always terminate within the synchronous execution instant in which they are called.

A number of functions are predefined under the form of operators over the predefined Boolean and numerical types.

function-name:
    function-ident
    predefined-operator

predefined-operator: one of
    and   or   not   <   >   <=   >=   <>   =   +   -   *   /   mod

B.2.6 Tasks

A task is a name for a potentially multiple-cycle routine defined in the host language. Like a procedure, it has zero or more pass-by-value arguments and zero or more pass-by-reference arguments.

When invoked with the **exec** statement, a task starts and may execute for zero or more instants. Special interface signals are used to notify the program of the task termination. See Appendix A.4 for more information about tasks.

B.3 Expressions

B.3.1 Data Expressions

Data expressions are formed from data atoms using functions and predefined operators. Type-checking is strict, and no implicit conversion is performed between the different types.

The atomic data expressions are the literals, the constants, the variables, the current (**?S**) and previous (**pre(?S)**) values of valued signals and sensors, the values of valued traps (**??T**), and the functions with no arguments. Recall that accessing an undefined signal or trap value is an error, but the compiler is unable to check it.

Complex expressions are formed by combining functions, the predefined operators, and parentheses "()." Parentheses are not always necessary because of the operator precedence rules defined in Section 2.5.1.

data-expr:
    identifier
    string-literal
    unsigned-literal
    boolean-literal
    (*data-expr*)
    ? *signal-ident*
    pre (? *signal-ident*)
    ?? *trap-ident*
    predefined-operator data-expr
    data-expr predefined-operator data-expr
    function-ident (*data-expr-list$_{opt}$*)

data-expr-list:
    data-expr
    data-expr-list , *data-expr*

B.3.2 Constant Atoms

When named constants are defined in Esterel, they can be assigned a value. To define such values, called constant atoms, Esterel largely restricts the syntax of data expressions.

constant-atom:
    constant-ident
    predefined-constant
    string-literal
    −$_{opt}$ *unsigned-literal*

B.3.3 Signal Expressions

Signal expressions are presented in Appendix A.1 along with their limitations in existing compilers.

signal-expr:
>*signal-ident*
>(*signal-expr*)
>**not** *signal-expr*
>*signal-expr* **and** *signal-expr*
>*signal-expr* **or** *signal-expr*
>*pre-expr*

pre-expr:
>**pre** (*signal-expr*)

B.3.4 Delay Expressions

Delay expressions are presented in Appendix A.1.

delay-expr:
>*delay-event*
>**immediate** *delay-event*
>*data-expr delay-event*

delay-event:
>*signal-ident*
>*pre-expr*
>[*signal-expr*]

delay-expr:
>*signal-expr*
>**immediate** *signal-expr*
>*data-expr signal-expr*

B.3.5 Trap Expressions

Trap expressions are used in conjunction with concurrent traps, as defined in Appendix A.2. Trap expressions allow the execution of a handler when a combinational expression of the concurrent trap codes is true.

trap-expr:
>*trap-ident*
>(*trap-expr*)
>**not** *trap-expr*
>*trap-expr* **and** *trap-expr*
>*trap-expr* **or** *trap-expr*

B.4 Statements

statement:
    control-flow-operator
    abort
    await
    procedure-call
    do-upto   *(deprecated)*
    do-watching   *(deprecated)*
    emit
    every-do
    exec
    exit
    halt
    if-test
    loop
    loop-each
    nothing
    pause
    present
    repeat
    run
    signal-decl-statement
    suspend
    sustain
    trap
    var-decl-statement
    weak-abort

control-flow-operator:
    sequence
    parallel
    block
    assignment

B.4.1 Control Flow Operators

sequence:
    statement ; *statement*
    sequence ; *statement*

parallel:
    statement || *statement*
    parallel || *statement*

block:
> [*statement*]

assignment:
> *var-ident* := *data-expr*

The sequence and parallel operators are associative, so their multi-way versions can be defined by expansion using the two-way versions and the parenthesis. The sequence operator ';' binds tighter than '||'. Therefore, the brackets are unnecessary in "[*p* ; *q*] || *r*."

B.4.2 abort: Strong Preemption

abort:
> abort *statement* when *delay-expr*
> abort *statement* when *abort-instance end-abort*
> abort *statement* when *abort-case-list end-abort*

abort-instance:
> *delay-expr*
> *delay-expr* do *statement*

end-abort:
> end abort$_{opt}$

abort-case-list:
> *abort-case-list$_{opt}$ abort-case*

abort-case:
> case *abort-instance*

The simplest form of the strong preemption statement is

> abort *p* when *d*

When the delay expression *d* is not immediate, the meaning of the statement is given by

```
trap T in
  suspend p when d ; exit T
||
  await d ; exit T
end
```

As its expansion shows, abort terminates either when *p* itself terminates, or when *d* becomes true and *p* is preempted. Upon preemption, control does not enter the statement *p*, the statement being suspended.

When *d*="immediate *s*," where *s* is a signal expression, control can be preempted during the start instant, before it enters *p*. In this case, the expansion is

```
present s else
   abort p when s
end
```

Like `await`, an extended form of the statement allows code to be executed upon preemption

```
abort p when d do q end
```

This is a short-hand for:

```
abort p when d ;
present s then q end
```

where s is the signal expression part of the delay expression d.

The multi-way form of `abort` allows us to give multiple preemption conditions. The form of the statement is

```
abort p when
   case d₁ do q₁
   ...
   case dₙ do qₙ
end
```

The statement p is preempted when at least one of the triggers is satisfied, and the corresponding q_i is started. If several triggers become true at the same time, the first one in the list activates the corresponding q_i. The multi-way `abort` statement is a short-hand for

```
abort
   ...
      abort
         p
      when dₙ do qₙ end
   ...
when d₁ do q₁ end
```

B.4.3 `await`: Strong Preemption

await:
```
await delay-expr
await abort-instance end-await
await abort-case-list end-await
```

end-await:
```
end await_opt
```

The `await` statement waits for a specified delay, then either terminates or runs the statement after the appropriate `do` clause. Its simplest form is await d

When $d=$"n s," with n a data expression and s a signal expression, the semantic expansion of the statement is

    suspend
      repeat n times pause end
    when [not s]

The delay expires in the instant where s becomes true for the nth time in instants following the start of the statement (the start instant is not counted). When d has no counter, the expansion is that of "await 1 d."

When $d=$"immediate s," with s a signal expression, the statement can terminate during the start instant. Its expansion is

    present s else await s end

The extended form of the statement allows code to be executed upon completion

    await d do q end

Its expansion is

    await d ;
    present s then q end

where we assumed that s is the signal expression part of d.

The multi-way form of await allows waiting for the first of a set of events. The form of the statement is

    await
    case d_1 do q_1
    ...
    case d_n do q_n
    end

The statement terminates when at least one of the triggers is satisfied. Then, the corresponding q_i is started. If several triggers become true at the same time, the corresponding q_i of lesser index is started. We give the semantic expansion of the multi-way await using the multi-way abort:

    abort
      halt
    case d_1 do q_1
    ...
    case d_n do q_n
    end

B.4.4 call: Procedure Call

procedure-call:
    call *proc-ident* (*var-ident-list*)(*data-expr-list*)

A procedure call awaits the computation of all the valued signals (shared variables) involved in the data expression list. Then, it evaluates the data expressions provided as value parameters, it performs its computation and stores the results in the sequential variables provided as reference parameters.

B.4.5 do-upto: Conditional Iteration (deprecated)

do-upto:
    do *statement* upto *delay-expr*

The expansion of "do p upto d" is "abort p ; halt when d."

B.4.6 do-watching: Strong Preemption (deprecated)

do-watching:
    do *statement* watching *delay-expr* *do-watching-end*$_{opt}$

do-watching-end:
    timeout *statement* end timeout$_{opt}$

This is a deprecated form of the **abort** statement. Its general form is

    do p watching d timeout q end

equivalent to

    abort p when d do q end

The short version, without a handler statement q, expands into the short version of the **abort** statement.

B.4.7 emit: Signal Emission

emit:
    emit *signal-ident*
    emit *signal-ident* (*data-expr*)

The semantic expansion of the valued signal emission along with pertinent semantic considerations is presented on page 23.

B.4.8 every-do: Conditional Iteration

every-do:
    every *delay-expr* **do** *statement* **end every**$_{opt}$

Similar to **loop...every**, with the exception that it awaits the trigger event in order to start the body the first time. Moreover, since the body cannot instantly restart upon normal termination, immediate triggers can be used.

Consider the general form of the statement "**every** d **do** p **end**." When d is not an immediate *delay-expr*, the semantic expansion is

    await d ;
    loop p each d

When $t=$"**immediate** s," the expansion is

    await d ;
    loop p each s

B.4.9 exec: Task Execution

exec:
    exec *task-instance* *end-exec*
    exec *exec-case-list* *end-exec*

task-instance:
    task-ident (*var-ident-list*) (*data-expr-list*) *return-handle*

exec-case-list:
    exec-case-list$_{opt}$ *exec-case*

exec-case:
    case *task-instance*

return-handle:
    return *signal-ident*
    return *signal-ident* **do** *statement*

exec-end:
    end exec$_{opt}$

Tasks and the **exec** statement are described, along with their semantics, in Appendix A.4.

B.4.10 exit: Trap Exit

exit:
>   exit *trap-ident*
>   exit *trap-ident* (*data-expr*)

Traps and the `exit` statement are described, along with their semantics, in Appendix A.2.

B.4.11 halt: Wait Forever

halt:
>   halt

Waits forever (until preempted). Its expansion is

```
loop pause end
```

B.4.12 if: Conditional for Data

if-test:
>   if *data-expr* *then-part*$_{opt}$ *elsif-part-list*$_{opt}$ *else-part*$_{opt}$ end if$_{opt}$

elsif-part-list:
>   *elsif-part-list*$_{opt}$ *elsif-part*

elsif-part:
>   elsif *data-expr* *then-part*$_{opt}$

then-part:
>   then *statement*

else-part:
>   else *statement*

Data test. The simplest form of the statement is

```
if data-expr then p else q end
```

Either the `then` or `else` branch may be missing. A missing branch is equivalent to `nothing` as body. The kernel expansion of the statement is

```
var v:Boolean in
  v:=data-expr ;
  if v then p else q end
end
```

This assumes that v does not cause a naming conflict.

The general form of the statement is

```
if data-expr₁ then p₁
elsif data-expr₂ then p₂
...
elsif data-exprₖ then pₖ
else pₖ₊₁
end
```

Its semantic expansion is

```
if data-expr₁ then p₁ else
  if data-expr₂ then p₂ else
    ...
      if data-exprₖ then pₖ else pₖ₊₁ end
    ...
  end
end
```

B.4.13 *loop*: Infinite Loop

loop:
    loop *statement* end loop$_{opt}$

Loop the body indefinitely. The body statement p must be statically non-instantaneous (cf. Section 4.6).

B.4.14 *loop-each*: Conditional Iteration

loop-each:
    loop *statement* each *delay-expr*

The semantic expansion of "loop p each t" is

```
loop
  abort
    p ; halt
  when t
end
```

When started, the statement starts p. Then, each time t becomes true, the body is preempted and instantly restarted. If p terminates its execution before preemption, the control is retained by the halt statement. To ensure the absence of causality cycles, the trigger must not be instantaneous (current compilers ignore the immediate keyword).

B.4.15 *nothing*: No Operation

nothing:
>   nothing

Instantly passes control in sequence.

B.4.16 *pause*: Unit Delay

pause:
>   pause

Retain control until the next reaction.

B.4.17 *present*: Conditional for Signals

present:
>   present *present-event* *then-part$_{opt}$* *else-part$_{opt}$* end present$_{opt}$
>   present *present-case-list* *else-part$_{opt}$* end present$_{opt}$

present-event:
>   *signal-expr*
>   [*signal-expr*]

present-case-list:
>   *present-case-list$_{opt}$* *present-case*

present-case:
>   case *present-event*
>   case *present-event* do *statement*

The simplest form of the statement is

>   present *signal-expr* then p else q end

One (or both) of the branches can be absent. A missing branch is equivalent to **nothing**. Note that the statement with no branch does have an influence through constructive causality constraints, i.e., it awaits the moment where the present/absent status of the test expression is known.

The multi-way form of the **present** statement is

```
present
case signal-expr₁ do p₁
case signal-expr₂ do p₂
...
case signal-exprₖ do pₖ
else q
end
```

The else branch may be omitted. The order of cases is important, as shown by the semantic expansion:

    present *signal-expr*$_1$ then p_1 else
      present *signal-expr*$_2$ then p_2 else
        ...
        present *signal-expr*$_k$ then p_k else q end
        ...
      end
    end

The **present** and **if** statements are semantically similar. They have been merged in the Esterel V7 language.

B.4.18   *repeat: Iterate a Fixed Number of Times*

repeat:
    positive$_{opt}$ **repeat** *data-expr* **times** *statement* **end repeat**$_{opt}$

Repeat the body statement a fixed number of times, given by the value of the integer data expression. Like for the basic **loop** statement, the body must be statically non-instantaneous (cf. Section 4.6).

The simplest form of the statement is

    repeat *data-expr* **times** p **end**

In this case, the kernel expansion is:

```
trap T in
  var v:=data-expr:integer in
    loop
      if v>0 then v:=v-1 ; p else exit T end
    end
  end
end
```

A variant of the statement repeats its body at least once, even if the evaluation of the expression produces a negative or 0 value:

    positive repeat n **times** p **end**

In the kernel expansion, this amounts to adding the following statement just before the loop.

```
if v<=0 then v:=1 end ;
```

B.4.19   run: Module Instantiation

run:
>   **run** *name-renaming*
>   **copymodule** *name-renaming*   *(deprecated)*
>   **run** *name-renaming* [*renaming-list*]
>   **copymodule** *name-renaming* [*renaming-list*]   *(deprecated)*

name-renaming:
>   *module-ident*
>   *module-ident* / *module-ident*

renaming-list:
>   *renaming*
>   *renaming-list* ; *renaming*

renaming:
>   **type** *type-renaming-list*
>   **constant** *const-renaming-list*
>   **function** *function-renaming-list*
>   **procedure** *proc-renaming-list*
>   **task** *task-renaming-list*
>   **signal** *signal-renaming-list*

type-renaming-list:
>   *type-renaming*
>   *type-renaming-list* , *type-renaming*

type-renaming:
>   *type-name* / *type-ident*

const-renaming-list:
>   *const-renaming*
>   *const-renaming-list* , *const-renaming*

const-renaming:
>   *constant-atom* / *constant-ident*

function-renaming-list:
>   *function-renaming*
>   *function-renaming-list* , *function-renaming*

function-renaming:
>   *function-name* / *function-ident*

proc-renaming-list:
    proc-renaming
    proc-renaming-list , proc-renaming

proc-renaming:
    proc-ident / proc-ident

task-renaming-list:
    task-renaming
    task-renaming-list , task-renaming

task-renaming:
    task-ident / task-ident

signal-renaming-list:
    signal-renaming
    signal-renaming-list , signal-renaming

signal-renaming:
    signal-name / signal-ident

The semantics of the **run** statement is given in Section 2.5.3. The deprecated **copymodule** keyword is equivalent to **run**.

Note sub-modules can also be renamed, which is useful when debugging an application. Different instances of the same module can be given different names.

Renaming sections can be mixed freely in the renaming list of a **run** statement. Any renaming section must contain at least one renaming. The same object may not be renamed twice. No other restrictions are placed on the number or size of renaming sections.

B.4.20 `signal`: Local Signal Declaration

signal-decl-statement:
    signal *signal-decl-list* **in** *statement* **end signal**$_{opt}$

signal-decl-list:
    signal-decl
    signal-decl-list , signal-decl

signal-decl:
    signal-ident channel-description$_{opt}$

channel-description:
    : *channel-type*

(*channel-type*)
:= *data-expr* : *channel-type*

channel-type:
    type-name
    combine *type-name* **with** *function-name*

The local signal declaration statement is described and decomposed into kernel statements in Section 2.2. Here, we only provide some clarifications.

The first two definitions of *channel-description* are semantically equivalent; the version with : is preferred.

The declaration of multiple signals is done with

    signal *signal-decl$_1$*,..., *signal-decl$_k$* **in**
      p
    end

which expands into

    signal *signal-decl$_1$* **in**
      ...
      signal *signal-decl$_k$* **in**
        p
      end
      ...
    end

B.4.21 **suspend**: *Preemption with State Freeze*

suspend:
    suspend *statement* **when** *delay-expr*

The primitive form of the **suspend** statement, defined in Section 2.2, is the fundamental strong preemption primitive from which all other strong preemption statements are derived.

In the general form of the statement, the single signal can be replaced with a general delay expression d:

    suspend *p* **when** *d*

When d is a *delay-event*, the semantics of **suspend** are given by the following expansion, which can be further transformed into a kernel statement according to the expansion of signal expressions given in Appendix A.1.

```
signal Aux in
  suspend p when Aux
||
  loop present d then emit Aux end ; pause end
end
```

When $d=$"immediate s," where s is a *delay-event*, the body of **suspend** is only started when s becomes false. The expansion is

```
await immediate [not s] ;
suspend p when s
```

Counted delay expressions should not be used with **suspend**. However, compilers such as the INRIA Esterel V5 compiler accept them.

B.4.22 *sustain*: Emit a Signal Indefinitely

sustain:
      sustain *signal-ident*
      sustain *signal-ident* (*data-expr*)

When started, the **sustain** statement emits its signal at all instants until it is preempted. The signal is not emitted at instants where the statement is suspended. The kernel expansion of "**sustain S**" is

```
loop emit S ; pause end
```

B.4.23 *trap*: Trap Declaration and Handling

trap:
      trap *trap-decl-list* **in** *statement trap-handler-list* **end trap**$_{opt}$

trap-decl-list:
      trap-decl
      trap-decl-list , *trap-decl*

trap-decl:
      trap-ident channel-description$_{opt}$

trap-handler-list:
      trap-handler-list$_{opt}$ *trap-handler*

trap-handler:
      handle *trap-expr* **do** *statement*

Traps and the trap declaration statement are described in Appendix A.2.

B.4.24 *var*: Local Variable Declaration

var-decl-statement:
      var *var-decl-list* **in** *statement* **end var**$_{opt}$

var-decl-list:

var-decl
    var-decl-list , *var-decl*

var-decl:
    var-init-list : *type-ident*

var-init-list:
    var-init
    var-init-list , *var-init*

var-init:
    var-ident
    var-ident := *data-expr*

The local variable declaration statement is described in Section 2.2. The semantic expansion of multiple variable declaration is trivial.

B.4.25 weak abort: Weak Preemption

weak-abort:
    weak abort *statement* **when** *delay-expr*
    weak abort *statement* **when** *abort-instance end-weak-abort*
    weak abort *statement* **when** *abort-case-list end-weak-abort*

end-weak-abort:
    end abort$_{opt}$
    end weak abort

Weak preemption is syntactically similar to **abort**. It allows the body statement to perform its computation during the execution instant where it is preempted. Its simplest form is

    weak abort *p* **when** *d*

Regardless of the form of the trigger *d*, the semantic expansion of the statement is

```
trap T in
  p ; exit T
||
  await d ; exit T
end
```

The extended form of the statement allows code to be executed upon preemption

    weak abort *p* **when** *d* **do** *q* **end**

In this form, the expansion is

> weak abort p when d;
> present s then q end

where s is the *signal-expr* part of d.

The multi-way form of **weak abort** allows multiple preemption conditions. The form of the statement is

> weak abort p when
>   case d_1 do q_1
>   ...
>   case d_n do q_n
> end

Like the multi-way **abort**, the weak version activates only one of the handlers, as shown by its expansion.

> signal $T_1, \ldots, T_n$ in
>   weak abort
> p
> ||
>   await d_1 do emit T_1 end
> ||
>   ...
> ||
>   await d_n do emit T_n end
>   when [T_1 or...or T_n] do
>   present T_1 then q_1 else
>   ...
>   present T_n then q_n end
>   ...
>   end
>   end
> end

B.5 Modules

module:
>   **module** *module-ident* : *module-interface$_{opt}$* *statement* *end-module*

end-module:
>   **end module**
>   **.**   *(deprecated)*

module-interface:

module-interface$_{opt}$ interface-decl ;

In Esterel, the basic programming unit is the module. The description of modules is in Section 2.5. Below, we give the syntax of the interface declarations.

B.5.1 Interface Declarations

interface-decl:
    interface-signal-decl
    sensor-decl
    relation-decl
    type-decl
    constant-decl
    function-decl
    procedure-decl
    task-decl

Signals and Sensors

interface-signal-decl:
    interface-signal-type signal-decl-list

interface-signal-type:
    `input`
    `output`
    `inputoutput`
    `return`

sensor-decl:
    `sensor` *sensor-list*

sensor-list:
    sensor
    sensor , *sensor*

sensor:
    signal-ident : *type-name*
    signal-ident (*type-name*)

Relations

relation-decl:
    relation *relation-list*

relation-list:
    relation
    relation-list , *relation*

relation:
    signal-ident => *signal-ident*
    incompatibility-list

incompatibility-list:
    signal-ident # *signal-ident*
    incompatibility-list # *signal-ident*

Types

type-decl:
    type *type-ident-list*

type-ident-list:
    type-ident
    type-ident-list , *type-ident*

Constants

constant-decl:
    constant *one-type-constant-decl-list*

one-type-constant-decl-list:
    one-type-constant-decl
    one-type-constant-decl-list , *one-type-constant-decl*

one-type-constant-decl:
    constant-list : *type-ident*

constant-list:
    constant
    constant , *constant-list*

constant:
    constant-ident = *constant-atom*

constant-ident

The first form of the last rule binds a literal or other constant to the given constant identifier. The second form only defines the name and type of the constant; the generated code simply uses the identifier verbatim, assuming the value will be defined in the host language.

Functions

function-decl:
> **function** *function-list*

function-list:
> *function*
> *function-list* , *function*

function:
> *function-ident* (*ident-list*) : *type-name*

A function declaration specifies the function name, its parameter list, and its return type.

Procedures

procedure-decl:
> **procedure** *procedure-list*

procedure-list:
> *procedure*
> *procedure-list* , *procedure*

procedure:
> *proc-ident* (*ident-list*) (*ident-list*)

Procedure declarations specify the procedure name and two formal parameter lists. The first list contains the types of parameters that are passed by reference (i.e., that can be modified by the procedure). The second contains the types of parameters passed by value.

Tasks

task-decl:
> **task** *task-list*

task-list:
> *task*

task-list , *task*

task:
task-ident (*ident-list*) (*ident-list*)

C

The C Language Interface*

The INRIA compiler system defined a C language interface that has become the de facto standard for generated code, providing a way to use different Esterel compilers without modifying the run-time environment.

The main object in the generated code is the reaction function, which inherits the name of the compiled Esterel module. Inputs and outputs are communicated through input and output functions whose names are automatically computed from the module and signal names. The run-time interface is purely procedural. No assumption is made about any operating system interface.

The interface is designed to allow a user to provide inputs at the beginning of a reaction and call the reaction function, which calls user-defined output functions as a result. Such an interface obviously does not make assumptions about sources of inputs or the destination of outputs; this makes it possible to use the generated code in fairly arbitrary environments.

As explained in Section 1.3, the generated C code may require user-supplied code to define the types, constants, functions, procedures, and tasks used in the module body. This is called data-handling code. In addition, an execution shell is necessary to interface at run-time with the outside world, i.e., to detect input events, call the reaction function, and perform output actions.

From this point on, we assume that the generated C code of an Esterel program named PROG has been placed in PROG.c.

C.1 Overview

If the Esterel source program refers to user-defined types, constants, functions, procedures, or tasks, the user must link the generated code with some data-handling code that defines the implementation of these objects.

---

*Much of this chapter is from the manual of the INRIA compiler, by Berry et al. [13].

Definitions of user-defined types must be supplied when compiling the generated `PROG.c` file. These type definitions must appear in a file `PROG.h`, which is automatically `#included` by `PROG.c`. In addition to type definitions, the `PROG.h` file can contain inline definitions of constants, functions, and procedures by `#define` directives. Constants, functions and procedures not defined this way can appear in another C file, typically named `PROG_data.c`.

For execution, the generated code must also be linked with the execution shell that realizes the interface with the outside world, i.e., detects input events, calls the reaction function, and realizes output events. The way the execution shell performs these actions is implementation- and user-dependent.

If the user wishes the `PROG` module to react to an input event composed of two simultaneous input signals `I1` and `I2`, where `I2` is a valued signal to be passed the value given by the C expression `exp`, the user first calls the two automatically generated input C functions `PROG_I_I1` and `PROG_I_I2` in any order

```
PROG_I_I1();
PROG_I_I2(exp);
```

then calls the reaction function:

```
PROG();
```

Notice that absence is the default. As in Esterel, nothing needs to be called to indicate an absent signal.

Reactions are not reentrant and must be executed atomically. In particular, during the execution of the reaction function, neither user input C functions nor the reaction function itself should be called.

During its execution, the reaction function may call user-supplied C data-handling functions that implement the functions, procedures, and tasks declared in the Esterel program. It may also read a sensor `S1` by calling the user-supplied sensor C function `PROG_S_S1`, which should return the sensor's value.

If the Esterel program emits the output signal `O1`, the generated C code calls the user-supplied output C function `PROG_O_O1`. If `O1` is valued, the value is passed as the lone argument to `PROG_O_O1`.

To summarize, the user must write functions to read sensors, named `PROG_S_xx`, and output functions, named `PROG_O_xx`. The reaction function `PROG` and the input functions `PROG_I_xx` are generated by the Esterel compiler. All functions related to program input, output, or execution are prefixed with the program name, but the data-handling function names and the module names are not. Therefore, C reserved words (e.g., `switch`, `while`) and standard library function names (e.g., `printf`) should not be used as Esterel function or module names.

C.2 C Code for Data Handling

C.2.1 Defining Data-handling Objects

If `PROG` is the name of the Esterel module, then a directive of the form

    #include "PROG.h"

is included in the `PROG.c` file if the Esterel input file declares a type, a constant, a function, a procedure, or a task.

The `PROG.h` file must contain the code needed to compile the generated `PROG.c` C file separately. Therefore, `PROG.h` must at least contain the C definition of the user-defined types used in the source program. It can also contain inline constant, function, and procedure definitions by `#define` directives.

The constants, functions, and procedures used in the Esterel source program but not `#defined` in `PROG.h` are automatically declared `extern` in `PROG.c` and can be defined in other C files.

The C names of types, constants, functions, and procedures must match those in the Esterel file.

C.2.2 Predefined Types

The basic type `integer` is implemented as `int`. The basic type `boolean` is also implemented as `int`, with constants `false` = 0 and `true` = 1. The basic types `float` and `double` are implemented by their eponymous C types.

There are some peculiarities for the basic type `string`, since there is no real string type in C. It is implemented as follows: the type itself is declared as `char*`; a variable `VAR` of type `string` is declared as an array of characters and is allocated in the generated code by the declaration

    char __PROG_Vxx[STRLEN];

where `xx` is some allocation number. In `PROG.c` `STRLEN` is defined with

    #define STRLEN 81

which may be modified or overridden by a compilation command of the form

    cc -DSTRLEN=125 -c PROG.c

String assignment is done by copy using the C `strcpy` function.

C.2.3 User-defined Types

The file `PROG.h` must contain a type definition for each user-declared type in the source program. The C names must match the names in the Esterel file. Any declaration of a variable or signal of type `T` in the source program generates a C declaration of the form

```
T __PROG_Vxx;
```

The C `typedef` construct is intended for declaring types. This is compulsory for structures. For example,

```
typedef struct {
    int hours;
    int minutes;
    int seconds;
} TIME;
```

Assignment Functions

If `T` is a user-defined type, the declaring module can make use of the predefined assignment, equality, and inequality operators for objects of type `T`. When one of these operators is used, the user must supply a C definition for it, as follows.

A call to the assignment operator is generated when

- there is an explicit assignment or variable initialization of type `T`, i.e., the assignment symbol ":=" is used somewhere for a variable of type `T`;

- there is a valued signal of type `T`, either in the main module or in one of its submodules (emitting valued signals requires assignment); or

- an `exec` statement calls a task with a reference parameter of type `T`.

If at least one of the above conditions is met, the user must write an assignment function for the user-defined type `T`. This C function takes two arguments, the first one of type `T*`, the second one of type `T`. The return value of the function should be `void`. For example, the source Esterel assignment

```
X := exp
```

where `X` and `exp` are of type `T` generates a call of the form

```
_T(&__PROG_Vxx, exp)
```

In the case of structured data types, it is recommended that assignment functions perform a complete (recursive) copy of the source. Strange behaviors can appear otherwise. If C natively supports assignment for type `T`, then `_T` can be simply defined as

```
#define _T(x,y) (*(x)=y)
```

As usual, the parentheses avoid operator precedence errors. With this macro, an assignment of the form `X := exp` becomes

```
(*(&__PROG_Vxx) = exp)
```

which is interpreted as `__PROG_Vxx = exp` by the C compiler.

Equality Functions

A reference to the equality function of type T is generated when at least a comparison "=" between objects of type T appears in the Esterel source.

In this case, an equality function named _eq_T must be supplied. This C function is passed two arguments of type T and is expected to return an int. For example, the definition of equality for type TIME could be

```
int _eq_TIME(TIME t1, TIME t2)
{
   return t1.hours   == t2.hours &&
          t1.minutes == t2.minutes &&
          t1.seconds == t2.seconds;
}
```

Similarly, a reference to the inequality function of type T is generated if and only if a comparison "<>" between objects of type T appears in the module or one of its submodules. If a comparison is used, an inequality function named _ne_T must be defined. This C function has the same type as _eq_T and may be defined as the negation of _eq_type.

An Array Example

Consider the following example involving an array encapsulated in a struct. Note the use of the assignment and inequality C functions of the underlying TIME type (described above).

```
typedef struct {
  TIME array[10];
} ARRAY_10_OF_TIME;

/* assignment */
_ARRAY_10_OF_TIME(ARRAY_10_OF_TIME *pa, ARRAY_10_OF_TIME a)
{
 int i;
 for (i = 0; i < 10; i++)
    _TIME(&(pa->array[i]), a.array[i]);
}

/* equality */
int _eq_ARRAY_10_OF_TIME(ARRAY_10_OF_TIME a1,
                        ARRAY_10_OF_TIME a2)
{
 int i;
 for (i = 0; i < 10; i++)
    if (_ne_TIME(a1.array[i], a2.array[i])) return 0;
```

```
    return 1;
  }

  /* inequality */
  int _ne_ARRAY_10_OF_TIME(ARRAY_10_OF_TIME a1,
                           ARRAY_10_OF_TIME a2)
  {
    int i;
    for (i = 0; i < 10; i++)
      if (_ne_TIME(a1.array[i], a2.array[i])) return 1;
    return 0;
  }
```

C.2.4 Constants

Each constant used but not initialized in the Esterel program must be defined with the same name in C. A constant can be defined either by a `#define` directive in PROG.h or by a standard C variable definition. If not `#defined`, a constant is automatically declared **extern** in PROG.c. It can therefore be defined in any other file. Consider the example

```
constant NUMBER_OF_PERSONS : integer,
         LUNCH_TIME : TIME;
```

Then PROG.h could contain

```
#define NUMBER_OF_PERSONS 45
```

and PROG_data.c could contain

```
TIME LUNCH_TIME = { 12, 0, 0 };
```

C.2.5 Functions

Each function called by the Esterel program must be defined in C as either a macro (a `#define` directive in PROG.h) or as a C function. If not `#defined`, the C function is automatically declared **extern** in the generated file PROG.c.

The type of a C function must match its type in the Esterel code. For example, consider defining the Esterel function declared as

```
function FETCH(ARRAY_10_OF_TIME, integer) : TIME;
```

A possible definition is

```
TIME FETCH(ARRAY_10_OF_TIME a, int i)
{
   return a.array[i];
}
```

C.2.6 Procedures

Each procedure called in the Esterel program must be defined either as a macro (a #define directive in PROG.h) or as a C function. If not #defined, the C function is automatically declared to be extern in PROG.c.

An Esterel procedure has two argument lists: the first contains pass-by-reference arguments; the second has of pass-by-value arguments. In C, the two lists are concatenated into a single list of arguments. The reference arguments are passed by pointers and the value arguments are passed by value. For example, for the Esterel procedure declaration

```
procedure PROC(T1)(T2);
```

the corresponding C function PROC has two arguments: a pointer to type T1 and an object of type T2. A consistent C declaration is

```
void PROC(T1 *pt1, T2 t2)
{
    /* ... */
}
```

Here is another example.

```
procedure STORE(ARRAY_10_OF_TIME)(integer, TIME);
```

can be declared in C as

```
void STORE(ARRAY_10_OF_TIME *pa, int i, TIME t)
{
    _TIME(&(pa->array[i]), t);
}
```

Note the use of the assignment function _TIME.

There is an exception for strings. Since they are already pointers, pass-by-reference arguments are passed the same as pass-by-value. For example,

```
procedure STORE_CHAR(string)(integer, CHAR);
```

can be implemented by

```
void STORE_CHAR(char *s, int i, CHAR c)
{
    s[i] = c;
}
```

where CHAR is a user type implemented by char. The declaration "char *s" is used instead of "string *s," which would be the case for a user-defined type.

C.3 The Reaction Interface

The reaction function generated by the Esterel compiler for a program PROG is called PROG. The generated code also contains a function for each input and inputoutput signal; the user is responsible for providing a function or macro for each output, inputoutput, and sensor.

C.3.1 Input Signals

For each input signal IS, the Esterel compiler generates an input C function called PROG_I_IS, which takes an argument of the appropriate type if the signal IS conveys a value. For example, from the Esterel declarations

```
input WATCH_MODE_COMMAND;
input WATCH_TIME (WATCH_TIME_TYPE);
return R;
```

appearing in a module named DISPLAY, the compiler generates the following functions:

```
void DISPLAY_I_WATCH_MODE_COMMAND() { ... }
void DISPLAY_I_WATCH_TIME(WATCH_TIME_TYPE __V) { ... }
void DISPLAY_I_R() { ... }
```

The objective here is to match the execution mechanism presented in Section 1.3, Figure 1.7. When a program PROG should react to an input event composed of one or more input signals, the associated input C functions should be called before calling the reaction function PROG.

More precisely, the input event corresponding to the current call of the reaction function is formed of all the input signals XX whose input C functions PROG_I_XX have been called since the previous reaction function call.

The input function of an input signal can be called several times by the execution engine for a single execution instant. When the signal is pure (not valued), the first call sets the signal as present, and subsequent ones do nothing. When the signal is valued and combined, the values provided by the multiple calls are combined using the combine function. When the signal is valued but not combined, the last call to the input function sets the value of the signal for the current reaction (previous values are discarded).

The order in which input functions are called between two calls of the reaction function does not matter.

For example, assume that some signals have arrived from the external world, say a pure signal IS1 and an integer-valued signal IS2 conveying the integer value 3. To perform the corresponding program reaction, one must first call the two automatically generated input functions PROG_I_IS1 and PROG_I_IS2 and then call the C function PROG. One can execute the following sequence:

```
PROG_I_IS1();    /* input IS1 is present */
PROG_I_IS2(3);   /* input IS2 is present with value 3 */
PROG();          /* reaction function */
```

If IS is a combined integer signal, with "+" as its combine operator, then the execution of the following sequence

```
PROG_I_IS(1);
PROG_I_IS(2);
PROG();
```

corresponds to the execution of the Esterel program PROG in the context where IS is present with value $1 + 2 = 3$. This sequence is equivalent to the following:

```
PROG_I_IS(3);
PROG();
```

C.3.2 Return Signals

Return signals are particular input signals used to signal the completion of external tasks. In the generated C code, return signals are handled exactly as standard input signals.

C.3.3 Output Signals

For each output signal OS, the user must write a void output C function PROG_O_OS. When the signal OS is valued with type T, then PROG_O_OS must take an argument of type T. This function is automatically called by the reaction function PROG if the signal is emitted.

The order of the output function calls performed by the reaction function is arbitrary and unspecified. Assume that a reaction causes the output of a pure signal OS1 and of an integer signal OS2 with value 4. Then PROG calls the user-defined C functions PROG_O_OS1 and PROG_O_OS2 with the appropriate arguments; the following two calls will be executed in some unpredictable order in the body of PROG.

```
PROG_O_OS1();
PROG_O_OS2(4);
```

The user-provided C functions PROG_O_OS1 and PROG_O_OS2 are responsible for the communication with the actual environment.

C.3.4 Inputoutput Signals

For an inputoutput signal IOS, an input C function PROG_I_IOS is automatically generated as for an input signal, and the user must write an output C function PROG_O_IOS as for an output signal.

An inputoutput signal IOS received by the reactive program behaves as if it was emitted by it, and is therefore re-emitted outside whenever received in a reaction.

- If IOS is a pure inputoutput signal then PROG_O_IOS is called if IOS is received or emitted by the program. Therefore, PROG_O_IOS is always called by the reaction function if PROG_I_IOS was called before.

- If IOS is a non-combined signal, there are two cases:

    - If IOS is received by the program, it should not be emitted twice (either by the environment or the program) in the same reaction. The function PROG_O_IOS is called with argument the value received by PROG_I_IOS.

    - If IOS is not received, then PROG_O_IOS is called when IOS is emitted by the program. The argument of PROG_O_IOS is the emitted value.

- If IOS is a combined signal, then all the emitted or received values are combined using the signal's combine function. The output function PROG_O_IOS is called after all internal emissions have been executed, with the combined value as argument.

C.3.5 Sensors

The reaction function accesses the current value of sensors by calling a user-supplied function. Let SE be a sensor of type T. If the program needs the current value of SE to perform its reaction, it calls the argumentless user-supplied sensor C function PROG_S_SE. This must return a value of type T, which is the sensor current value. To ensure sensor value consistency, the program calls each sensor C function at most once in a reaction. Here is an example:

```
int PROG_S_TEMPERATURE()
{
  return measure() ; //calls a library function
}
```

C.3.6 Reaction and Reset

For the Esterel main module PROG, the Esterel compiler generates three void argumentless C functions:

- the reaction function PROG

- the signal reset function _PROG_reset_input, which resets the input event construction process, according to the internal encoding of the execution shell

- the program reset function PROG_reset, which resets the program by assigning the state variables their initial value

The reset function should be called before any reaction is performed, and it calls in turn the signal reset function. The last line of the reaction function is a call to the signal reset function, to allow the correct acquisition of the input event for the next reaction. To perform a program reaction, one calls the input C functions and then calls the reaction function, e.g.,

```
PROG_I_IS1();
PROG_I_IS2(3);
PROG();
```

Programs often contain instantaneous initial statements, such as signal emissions or variable initializations, to be performed during the first reaction. To perform them, it is often useful (but not mandatory) to generate a "blank" initial event by calling the reaction function once before calling any input C function. This is equivalent to running the Esterel program for one reaction with a void input event.

C.3.7 Notes

The relations between input signals specified in the source program are not checked when the reaction function is called. The code may behave strangely if called with inputs which do not satisfy the relations. However, the execution shell may enforce these relations, like in the debugging-oriented code generated by the INRIA compiler (option "-I").

The combine functions associated with combined signals must be commutative and associative. Otherwise, the results of signal combinations can be arbitrary since the combination order depends on the action schedule chosen by the compiler.

The reaction function is not reentrant and its execution must be atomic. Therefore, during a call to the reaction function, it is not legal to call it again, or to call input C functions. Arbitrarily strange behaviors can arise otherwise. In particular, interrupt handling routines should never call directly input C functions or the reaction function. They should instead fill event queues to be read when the reaction function call terminates. One can also mask interrupts during the reaction function execution.

Access to uninitialized variables and uninitialized signal values are not checked when the reaction function is called.

C.4 Task Handling*

In this section, we describe the interface for `exec` statements. This code concretely reflects at C level the abstract task interface and synchronization, as described in Appendix A.4. It is organized in two layers. The low-level layer is a direct interface to run-time C data structures that contain all the required information about the status of `exec` statements. The optional higher-level layer provides the user with a functional interface. The functional interface is fairly simple but inflexible. The low-level interface provides more control but is more detailed.

Recall from Appendix A.4 that although the task interface speaks of controlling asynchronously-running processes, it does not actually run any. Instead, it generates and receives events which it assumes the user uses to control, say, threads provided by an operating system. The actual mapping between asynchronous events and synchronous events is performed by the execution shell.

C.4.1 The Low-level Layer: ExecStatus

If the main module is called `PROG`, the following generated C function returns the number of `exec` statements in the compiled program:

```
int PROG_number_of_execs();
```

The following generated function returns the number of `exec` statements associated with a task named `TASK` in the compiled program:

```
int PROG_number_of_execs_of_TASK();
```

The ExecStatus Structure

Each `exec` statement, which is uniquely identified by its return signal, is associated with a C structure of type `__ExecStatus` that contains all relevant information about the `exec` status just after a reaction. This structure can be recovered in three ways:

By name: For each `exec` of return signal R, the generated C code contains a variable `PROG_exec_status_R` declared as

```
__ExecStatus PROG_exec_status_R = /* ... */;
```

By absolute number: The generated code declares an array of pointers to the `__ExecStatus` variables, of size `PROG_number_of_execs()`, which has one entry for each `exec` statement:

---

*Tasks are probably the least-used feature in Esterel. We include this description for completeness and note that we have not addressed the issue of compiling Esterel programs with tasks.

```
__ExecStatus *PROG_exec_status_array[] = /* ... */;
```

By relative number: For each task TASK, the generated code contains an array of pointers to the execStatus variables, with one entry for each exec of that task. The size of the array is given by the function PROG_number_of_execs_of_TASK():

```
__ExecStatus *PROG_exec_status_array_of_TASK[] = /* ... */ ;
```
Here is the definition of the __ExecStatus structure:

```
typedef struct {
    unsigned int start           : 1;
    unsigned int kill            : 1;
    unsigned int active          : 1;
    unsigned int suspended       : 1;
    unsigned int prev_active     : 1;
    unsigned int prev_suspended  : 1;
    unsigned int exec_index;
    unsigned int task_exec_index;
    void (*pStart)(); /* takes a function as argument */
    int  (*pRet)();   /* may take a value as argument */
} __ExecStatus;
```

The meanings of these fields is as follows.

start is 1 when the exec statement starts and is not immediately killed (otherwise, it is 0). In that case, a new instance of the task code should be started in the current instant. See below for how to recover the actual parameter values from the pStart field.

kill is 1 when the exec statement is killed in the current instant. Then, the currently running instance of the task should be killed; notice that kill can only be 1 if there is such a running instance.

active is 1 when the exec statement is active in the current instant. This means that the exec is started in the current instant or has been started before, has not yet been killed, and that the task code has not yet returned.

suspended is 1 when the exec statement is active and suspended in the current instant by an enclosing suspend statement.

prev_active is 1 when the exec was already active in the previous Esterel instant.

prev_suspended is 1 when the exec was suspended in the previous instant.

exec_index is an integer uniquely identifying the exec statement. This index ranges between 0 and $n-1$ if the Esterel program contains n exec statements after full submodule instantiation.

task_exec_index is an integer uniquely identifying the exec statement among those referring to the same task. This index ranges between 0 and $p-1$ if the Esterel program contains p exec statements for this task after full submodule instantiation.

pStart is an auxiliary function pointer to be used at start time, i.e., when start is 1.

pRet is a pointer to the return function PROG_I_R associated with the return signal, if the name of the main module is PROG and the name of the return signal is R (recall that a return signal is just an input signal).

The function pointed to by pStart takes a user-provided function as argument, and the reference and value arguments are passed to this user function with the same convention as for a procedure (reference arguments as pointers; value arguments as values). A typical use is

```
if (exec_status.start)
    (*exec_status.pStart)(my_start);
```

This will call the user-provided function my_start with arguments corresponding to arguments passed to the task at start time.

The user-provided function my_start should perform two actions: starting the task in the environment and saving the pointers to the reference arguments for their update at return time.

Calling the *pRet or PROG_I_R function in the execution shell amounts to emitting R to signal the Esterel program that the task is completed. If the return signal is valued, the return function takes its value as argument. The return function can be called either directly using its full name PROG_I_R or indirectly through the pRet pointer.

When the return function is called, the locations pointed by the pointers passed at start time for reference arguments must contain the values updated by the task.

Notice that there are redundancies between the fields of __ExecStatus. For example, prev_active and prev_suspended could be computed directly by the user. They were included because they are easy to compute within the Esterel program and convenient for the user.

Reincarnation of exec Statements

An exec statement can be killed and restarted in the same instant. For example, by executing

```
loop
    exec T()() return R
each I
```

Here, when I occurs, there may be two active occurrences of the task code that the user has to manage properly. The first one is the one being killed; the second is the one being started. There can be no more than two such occurrences.

Handling Reference Arguments

Consider an Esterel variable X implemented as a C variable of location X, and assume that X is passed by reference in an exec statement over task TSK.

At starting time, the contents of X are copied into another location L whose address is passed to the user-level task starting function my_start. During task execution, the code of the task may freely modify the contents of L. At return time, i.e., when PROG_I_R (or equivalently *pRet) and then PROG are called, the contents of L are automatically copied back to X.

This copy-restore mechanism is necessary because it is possible to kill exec statements: if reference arguments could be modified in place at location X before an exec is killed, the value of X would change in the Esterel program, which is forbidden by the Esterel semantics, as explained in Appendix A.4.

Update of reference arguments must be performed in place in the location L passed to the user starting function my_start; this is why these pointers should be saved by my_start. Actual update of X by L is triggered only when the reaction function PROG is called with return signal R present (and of course only if the exec statement is not killed by an enclosing abortion statement).

C.4.2 The Functional Interface to Tasks

We now describe the much simpler functional interface. The user must provide four C functions:

- A *user start* function to start the task. This function receives the reference and value parameters plus a pointer to the __ExecStatus record of the exec statement as the last parameter; this is useful to index process-id tables associated with asynchronously running operating systems tasks, using the exec index fields.

- A *kill function* that is called when a task is killed, with a pointer to the __ExecStatus structure as argument.

- A *suspend function* that is called when the task becomes suspended, i.e., is now suspended but was not suspended in the previous instant (suspended=1, prev_suspended=0). This function also receives a pointer to the __ExecStatus structure as argument.

- A *resume function* that is called when the task should resume, i.e., when it was suspended at previous instant and it is neither suspended nor killed in the current instant. This function also receives a pointer to the __ExecStatus structure as argument.

To use the functional interface, one simply has to write a call to a specific STD_EXEC library macro with the return signal name and the user functions as arguments, once for each **exec** and right after each call to the reaction function:

```
#include "exec_status.h"

my_start()   { /* ... */ }
my_kill()    { /* ... */ }
my_suspend() { /* ... */ }
my_resume()  { /* ... */ }

/* ...context that calls the reaction function... */
PROG();  /* perform a transition */
STD_EXEC(R1, PROG,
         my_start_1, my_kill_1,
         my_suspend_1, my_resume_1);
STD_EXEC(R2, PROG,
         my_start_2, my_kill_2,
         my_suspend_2, my_resume_2);
```

A special __DUMMY__ function can be used if a user function is not necessary, e.g., if there is no **suspend** statement in the Esterel program:

```
STD_EXEC(R2, PROG,
         my_start_2, my_kill_2, __DUMMY__, __DUMMY__);
```

Finally, one can also write

```
STD_EXEC_FOR_TASK(TASK, PROG,
                  my_start, my_kill,
                  my_suspend, my_resume);
```

This calls STD_EXEC for all return signals of task TASK.

D
Esterel V7

Esterel V7, which has been developed at Esterel Technologies since 2001, is an evolution of Esterel V5. It was originally developed in cooperation with Michael Kishinevsky at Intel Strategic Cad Lab Portland, USA [11]. The extensions were mostly developed for hardware circuit design, but the language can also be used for software design. The Esterel V7 language is open (not proprietary) and its Language Reference Manual [29] has been submitted for IEEE standardization.

Esterel V7 extends the original Esterel V5 language in many respects: support for data definition, mostly eliminating the need for host data types, functions, and procedures (which are still available if needed); support for arrays of any base type and number of dimensions; definition of extensible interfaces to separate concerns between interface and behavior specification; support for registered signals using **next** instead of **pre**, and for temporary valued signals that do not preserve their value over time; support for oracles to model non-determinism; extension of the **emit** and **sustain** statements to support rich forms of concurrent and conditional emissions, directly extending Lustre equations [38]; support for extended tests that freely mix signal status and values; support of a new "**weak suspend**" statement that directly models clock-gating in circuits and can be implemented as such; last but not least, support for multiclock circuit design. To handle all these new features, Esterel V7 adds three new unit kinds to the Esterel V5 basic module one: data, interface, and multiclock units.

Because the generation of datapath components is easy for software targets, compiling Esterel V7 to software code is not much different from compiling Esterel V5, except for the new "**weak suspend**" statement, which is reasonably simple to add. Therefore, the contents of this book are also applicable to compiling Esterel V7 into software. Compiling the control portion of Esterel V7 to hardware is also very similar, but data-handling does become a more delicate issue.

D.1 Data Support

D.1.1 Basic Data Types

The former **boolean** type is renamed **bool** as in most languages, with constants **true** or '1 and **false** or '0. The mux(b,c,y) operator takes a Boolean b and two values x and y of the same arbitrary type and returns x if b is true and y if b is false; mux is undefined if b is.

The **float**, **double**, and **string** types are as in Esterel V5. They are not meant to be synthesizable in hardware.

The main change is for integers, which now come in two flavors: **unsigned** and **signed**, each with a given range. For instance, **unsigned<23>** is the range [0..22], which has 23 elements, while **signed<23>** is the range [−23..22], which has 46 components. The abbreviations **unsigned<[M]>** and **signed<[M]>** respectively denote **unsigned<2**M>** and **signed<2**(M-1)>**, which both have 2^M elements and fit in M bits. The abbreviation **unsigned** and **signed** denote **unsigned<[32]>** and **signed<[32]>** respectively. Unlike Esterel V5, where declared types are always externally defined in the host language, one can locally define types using the **type** definition.

```
type Pixel = unsigned<[8]>;
```

The Esterel V5 externally defined types are now accessed through the **host** keyword.

```
host type Time;
```

The **signed** and **unsigned** primitive types support exact arithmetic. Given two operands, the type of the result is the least type that can accommodate all possible results. For instance, unsigned addition and signed multiplication have the following types:

```
unsigned<M> + unsigned<N> → unsigned<M+N-1>
signed<M> * signed<N>   → signed<M*N+1>
```

These types are determined as follows: for unsigned addition, the worst case is $(M − 1) + (N − 1) = (M + N − 1) − 1$. for signed multiplication, the worst case is $-M * -N = (MN + 1) - 1$. The full rules are given in the V7 reference manual [29].

D.1.2 Arrays

One can declare arrays and arrays of arrays of any type and any number of dimensions. For example, X : Pixel[12] is an array of 12 unsigned values in [0..255]; Y : Pixel[12][6] is a two-dimensional array of the same values.

Partial indexing is possible in expressions: with the above definitions above, Y[2] is an array of dimension 6. Slicing can be done at any dimension and multiple slices are allowed. For example, the slice Y[2..9][2] is an array of dimension 8, itself indexed from 0 to 7, while the multiple slice Y[2..7][3..5] is an array of dimension 6 by 3. For full slices, one can omit the bounds, as for the column slice Y[..][2].

All operators are extended pointwise to arrays, using square brackets. For instance, if X and Y are unsigned arrays of the same type and dimension, then "X [+] Y" is an array of the same dimension where each component is the sum of the corresponding X and Y components.

Named types are very useful for arrays and arrays of arrays:

```
type Byte = bool[8];
type Word = bool[32];
type Memory = Word[1000]
```

Array literals can be defined either directly, such as $\{1,1,1,0,1,0,1,0\}$, or using repetition factors, such as $\{2\{1\}, 3\{1,0\}\}$, that denotes the same array. Multidimensional constants can also be defined: an example is $\{\{0,0\}, \{1,1\}\}$ of type unsigned[2][2], which can also be written $\{\{2\{0\}\}, \{2\{1\}\}\}$. Parameters can occur in constant definitions, as for $\{\{N\{0\}\}, \{N\{1\}\}\}$ where N is an unsigned constant.

D.1.3 Generic Types

In Esterel V5, a type declared using the "type T" declaration can be either generic, i.e., substituted at module run time, or host, i.e., defined in the host language. The same holds for constants, functions, procedures, and tasks. In Esterel V7, data objects can be generic, defined, or host. Generic objects are declared using the generic keyword and host objects are declared with host:

```
generic constant WordWirdth : unsigned;
type Word = bool[WordWidth];
host type Time;
```

D.1.4 Bitvectors

Single-dimensional arrays of base type bool, such as bool[32], are called bitvectors. They are ubiquitous in hardware and protocol design. They support special operations such as compact constant declaration, shifting, and concatenation.

Bitvector constants can be declared either using the normal array constant notation, e.g., $\{'0,'1,'1\}$. They can also be declared in binary, e.g., 'b110, octal, e.g., 'o364, and hexadecimal, e.g., 'xA2F7. Notice in binary, the higher-order bit is given first, so 'b110 is the same as $\{'0,'1,'1\}$.

To decompose composite word structures, one can name bits and fields:

```
type Word = bool[32];
map Word {
  sign_bit[31],              // single bit
  low_half_word[0..15];      // low-order slice
  high_half_word[16..31];    // high-order slice
}
```

Map fields can overlap. For multiple decomposition, one can use several maps for a single bitvector type. One can also name a map, e.g., "`map Halves: Word {...}.`".

D.1.5 From Numbers to Bitvectors and Back

Since the binary number system is hardly the only interesting one in hardware design, bitvectors are not implicitly treated as numbers. Conversion is done by predefined or user-defined functions.

The predefined conversion function u2bin of type unsigned<[M]> → bool[M] writes an abstract unsigned number into a concrete bitvector, while bin2u of type bool[M] → unsigned<[M]> performs the reciprocal binary-system reading from a bitvector to a number. The functions u2gray and gray2u do the same for Gray code instead of binary, and the functions u2onehot and onehot2u do the same for one-hot encoding. For unsigned<3>, one-hot encoding is 0 → 0001, 1 → 0010, 2 → 0100, and 3 → 1000. The type of u2onehot is unsigned<M> → bool[M+1]. Note that bin2u(u2onehot(n)) is a good way to compute 2^n.

The user can also define its own encoding using auxiliary functions u2code and code2u that take the code name as an additional argument. The actual encoding and decoding functions should then be provided in the synthesis- or run-time library.

D.1.6 Data Units

Data declarations can be grouped into data units, which are extensible:

```
data D1 :
  type Byte = bool[8];
  type Word = bool[32];
end data

data D2 :
  extends D1;
  generic MemSize : unsigned;
  type Memory = Word[MemSize];
end data
```

The extends declaration imports all objects from D1 in D2. Data can also be directly declared and data units be extended in interface, module, and multiclock definitions.

D.2 Signals

Esterel V7 provides the user with new signal attributes that are important for efficient hardware synthesis. We briefly describe them here.

D.2.1 Value-only Signals

Esterel V5 signals can be either pured or valued, with an extra interface signal type called the sensor. Esterel V7 replaces sensors by value-only signals, which have no status. This a frequent case for hardware signals, for which the optional status is often called a valid bit. A value-only signal is declared using the **value** keyword:

        output MemData : value Word;

D.2.2 Temporary Signals

In Esterel V5, the value of a valued signal is persistent between instants. This is natural for software but often too costly for hardware because of register area and power consumption. In Esterel V7, persistence is still the default, but a signal can also be declared temporary using the **temp** keyword. In this case, its value is not saved between instants*. For temporary valued signals, the value is available when the status is *present*; it may be either the previous value or undefined at other instants. For statusless value-only signals, it is the user's responsibility to know when the value is available; for instance, this could be due to the presence of another signal or on some Boolean condition on other signal values. Here is an example:

        output MemData : temp value Word;

D.2.3 Registered Signals

In Esterel V5, one can only emit a signal S for the current instant and read its current and previous status S and **pre(?S)** and its current and previous values ?S and **pre(?S)** if valued. Esterel V7 adds registered signals, which can only be local or output. Registered signals are emitted for the next instant. They are very important for hardware design since they are (most often) outputs of registers, which implies much better electrical properties than default combinational signals, and in particular the ability to cut critical paths. Let R and RV be declared by

        output R : reg,
               RV : reg unsigned<[8]>;

---

*The **temp value** association is so frequent that it may become the default in the future.

R is pure registered. It is emitted using the statement "emit next R;" when emitted, its status becomes true only at next instant. The valued registered signal RV is emitted by "emit ?RV <= exp," which sets the status and value for the next instant. For both R and RV, emission has no impact on the current status and value expressions R, RV, and ?RV. The status and value emitted in an instant for the next instant can be read at emission instant using the expressions next(R), next(RV), and next(?RV).

D.2.4 Signal Initialization

Signals and signal arrays can be initialized at declaration time. Here are some examples:

```
input I : unsigned<16> init 0;
         //default value until I received
output O : bool[4] init 'b0101;
signal X : temp unsigned<16>[4] init 0,
         //0 for all components
       Y : unsigned<[32]>[5] init 0,1,2,3,4 in ... end
```

For a persistent signal such as O or Y above, initialization is performed at signal declaration time. For a temporary signal such as X above, initialization is performed in every instant the signal is alive.

D.2.5 Oracles

Oracles are new signals used to model and formally verify non-deterministic behavior. They are not meant for synthesis.

An oracle is declared in a signal declaration with oracle keyword instead of signal. Conceptually, an oracle is not under user control and can take any status and value in each instant. On can view it as a locally declared extra input. Here is a way to model a non-deterministic choice between two statements p and q using a pure oracle:

```
oracle ArbitraryChoice in
  if ArbitraryChoice then
    p
  else
    q
  end if
end oracle
```

Oracles are regular signals with a unique status and value at each instant. If several tests for the same oracle are evaluated in the same instant, they will all take the same branch. This helps in structuring models and controlling the amount of non-deterministic, which is key to formal verification success.

D.3 Interfaces

D.3.1 Interface Declaration

Interfaces group signal declarations for later reuse. Here are read and write interfaces for a memory:

```
interface ReadIntf :
  input Read;
  output DataOut : Word;
end interface

interface WriteIntf :
  input Write;
  input DataIn : Word;
end interface
```

The full memory interface can be built in two ways. The first way is to use interface extension, which simply imports all components of an interface into a bigger one:

```
interface MemIntf :
  extends ReadIntf;
  extends WriteIntf;
end interface
```

In this case, MemIntf has the four signals declared by ReadIntf and WriteIntf. The other structuring choice is to declare ports typed by interfaces:

```
interface MemPortIntf :
  port ReadPort : ReadIntf;
  port Write : WriteIntf;
end interface
```

The signals are now called ReadPort.Read, ReadPort.DataOut, WritePort.Write, and WritePort.DataIn.

D.3.2 Interfaces and Modules

The **extends** declaration imports an interface into a module:

```
module Memory :
  extends interface Intf; // or simply extends Intf
  output ReadError;
  ...
end module
```

Note that the `ReadError` signal is declared directly in `Memory`. The interface of the `Memory` module can then be used as an interface in another module:

```
module ParityMemory :
  extends interface Memory;
  ouput ParityError;
  ...
end module
```

D.3.3 Mirroring an Interface

The mirror "`mirror Intf`" of an interface `Intf` is the interface obtained by swapping the input and output direction of signals in `Intf`. For example, `Read` is an output of "`mirror MemIntf`," which is typically extended by the module that communicates with the memory.

D.3.4 Interface Refinement in Modules

Only directions, types, and `temp` features of signals can be declared in an interface. The `reg` and `init` features can only appear in modules. A signal declared in an interface extended by a module can be refined in the module using a `refine` declaration. Here is an example:

```
module Memory :
  extends MemoryIntf;
  refine DataOut : reg init '0;
  ...
end module
```

When extending `Memory`'s interface by "`extends interface Memory`," the refined features are discarded since they belong to the `Memory` module body and not to its interface.

D.4 Statements

D.4.1 Expressions and Tests

Esterel V7 drops the distinction between status and Boolean expressions, which can now be freely mixed. For instance, to test whether a signal `I` is present with a positive value, one writes "`I and (?I>0)`". Consequently, there is no more need for the distinction between `present` and `if` tests; only `if` remains.

D.4.2 Static Replication

Sstatic statement replication is a powerful new feature of EsterelV7. It copies code under the control of a static variable called an iterator. Here is a simple way to emit all elements of an array in succession, each after a number of ticks that corresponds to its position in the array, starting from 1:

```
for i < N dopar
    await i+1 times tick;
    emit X[i]
end for
```

A for-dopar clause acts as a parallel, which as usual terminates when all components have terminated. Therefore, the above for static loop terminates after N+1 steps if X is of size N.

Static for loops replicate arbitrary statements. They are particularly useful to replicate submodule calls. Assume a module M has two inputs I and J and an output O. Then one can generate one copy of M for each matching components of arrays A, B, and X by writing:

```
for i < N dopar
    run M [A[i] / I, B[i] / J, X[i] / O]
end for
```

D.4.3 Enhanced Emit and Sustain Statements

In Esterel V7, emit and sustain statements now drive concurrent comma-separated emissions, which take a new equational form borrowed from Lustre [38]. A pure signal emission is simply denoted by the signal name, as for Esterel V5; it may be conditioned by a Boolean expression acting as an equation right-hand-side, as for "X <= exp". A valued emission has the form "?S <= exp", where exp is an arbitrary expression of the same type as the signal. Pure and valued emissions can be conditioned by if clauses, with the possibility of defining a case list using the '|' symbol. Finally, equation blocks can themselves be conditioned by if statements to share conditions between equations. Here are examples.

```
sustain {X, Y}

emit {
  X <= I and (?I>0),
  Y if I and (?I>0)   // equivalent
}

emit {
  ?V <= 0 if ?I = 1   // cases taken in order
       | 1 if ?I < 5
       | 2,
  ?W <= ?V+2
}
```

```
sustain {
  if ?I>0 then // if-case also allowed
    X,
    ?Y <= 3
  else
    Z
  end
}
```

Static `for` loops can appear within `emit` and `sustain`:

```
emit {
  for i < 5 do
    X[i] <= (?S < i) if I
  end for
}
```

All equations are concurrent in such a `for` loop. This does not make it possible to propagate carries. The `seq` sequential variant of `emit` and `sustain` computes equations sequentially to allow carry propagation. Here is the definition of a binary adder on pure signal arrays:

```
main module Adder :
  constant N : unsigned = 4;
  input A[N], B[N];
  output S[N+1];
  signal C[N+1] in // C[i] carry out of stage i
                  // C[0] absent since not emitted
    sustain seq {
      for i < N doup
        S[i] <= A[i] xor B[i] xor C[i],
        C[i+1] <= (A[i] and B[i]) or
                  (B[i] and C[i]) or (C[i] and A[i])
      end for,
      S[N] <= C[N]
    }
  end signal
end module
```

Within a sequential emission, `do` must be replaced by `doup` or `dodown` to specify the order in which the indices are evaluated.

D.4.4 Explicit and Implicit Assertions

There are two kinds of verification assertions in Esterel V7: explicit assertions and implicit assertions. Both can be verified dynamically during simulation or statically by formal verification.

Explicitly named assertions are declared using the **assert** keyword, among signal equations in **emit** or **sustain** statements:

```
await A;
emit {
  A_received,
  assert Apos = ?A>0
};
pause;
abort
  sustain {
    waiting_for_B.
    assert Xpos = ?X>0 if X
  }
when B
```

A named assertion needs no other declaration that its definition within **emit** or **sustain**. Note that it can be duplicated by static **for** loops or by reincarnation of signal declarations or parallel statements. An assertion is only checked during the lifetime of its **emit** or **sustain** statement. Here, Apos is only checked at first instant, while Bpos is checked from second instant on until one instant before B occurs.

Unnamed explicit assertions are generated by **assert** arithmetic operations. For instance, for **exp** of unsigned type, **assert<M>(exp)** checks that the value of **exp** is less than M.

Other implicit assertions are generated by arithmetic operations (e.g., the quotient of a division is non-zero) and by indexing arrays (the index is withing the array bounds).

D.4.5 Weak Suspension

Suspension of a statement by a **suspend** statement is strong: when suspended, a statement does not receive the control and performs no action. A **weak suspend** milder form of suspension is available in Esterel V7. It was originally introduced by Schneider in his variant of Esterel called Quartz [62]. The **weak suspend** statement is syntactically similar to the **suspend** statement:

```
weak suspend
  p
when exp
```

As for **suspend**, the guard exp is not tested in the first instant, and the whole statement terminates when its body does. After the first instant, the guard is evaluated at each instant. If it is false, the body acts normally and the whole statement terminates, pauses, or exits a trap if its body does. If *exp* is true, the body *p* is also executed for the instant but does not update its internal state. The **weak suspend** statement pauses if *p* terminates or pauses (termination is discarded). In that case, in the next instant *p* is restarted from the same control points and with the same values for all local signals and variables it declares local. At any instant, if *p* exits a trap, that trap is propagated irrespective of SUSP's presence, and the **weak suspend** is weakly aborted. Here is a basic example:

```
module M :
  input SUSP;
  output {X, Y} : unsigned init 0;
  output Done;
  weak suspend
    signal S : unsigned init 0 in
      pause;
      emit ?X <= pre(?X) + 1;
      pause;
      emit {
        ?S <= pre(?S) +1,
        Y <= ?S
      }
    end signal
  when SUSP;
  emit Done;
end module
```

Here is an execution sequence:

```
;
% Output: none
SUSP;
% Output: X=1
SUSP;
% Output: X=2
;
% Output: X=3
SUSP;
% Output: Y=1
SUSP;
% Output: Y=1
;
% Output: Y=1 Done
```

Since the **weak suspend** body always runs, there is an output in every instant but the first. Control and local data state does not change and termination is preempted when SUSP is present. Therefore, control stays at

the first **pause** and "**emit X**" is executed in steps 2 and 3, until step 4, where
SUSP is absent. Since **Y** is declared outside the **weak suspend**, its value is not
subject to suspension and is changed at steps 2, 3, and 4 Similarly, control
stays at second **pause** and "**emit Y**" is executed until step 7 where the **weak
suspend** statement terminates. However, since **S** is declared within **weak
suspend**, its persistent value **pre(?S)** of **S** does not change and remains 0
at steps 5, 6, and 7. At this steps, the current value **?S** is 1 by the **?S <=
pre(?S) +1**. That current value does not survive the suspension instants.

This behavior may look awkward for a software programmer, but it is
natural for a hardware designer since it is exactly the effect of clock-gating
the internal registers of p by **not(SUSP)**. Clock gating is key to power savings
in circuits; **weak suspend** provides direct semantics for it.

There is also an immediate form, where the guard is also evaluated in the
first instant:

> **weak suspend**
> p
> **when immediate** exp

The immediate form can be derived from the default delay form using a
non-trivial macro-expansion described in the V7 reference manual [29].

D.4.6 Signal Connection by Module Instantiation

Esterel V7 takes module instantiation much more seriously than Esterel V5.
In Esterel V5, a **run** statement is purely syntactic: the text of the run sub-
module is written in place of the **run** statement, with syntactic replacement
of interface signals as specified in the renaming list. In Esterel V7, submodule
signals keep their individuality and are appropriately connected to caller signals. The behavior differs in several cases, and in particular in a suspension
context. Consider the following program:

```
module Sub :
  input SubI;
  output Y;
  sustain Y <= pre(SubI)
end module

module Main :
  input I, SUSP;
  output X, Y;
  {
    sustain X <= pre(I)
  ||
    suspend
      run Sub [ I / SubI ]
```

```
          when SUSP
      }
    end module
```

In Esterel V5, `I` in `Main` captures `SubI` in `Sub`, and `X` and `Y` always have equal status. In Esterel V7, `I` in `Main` and `SubI` in `Sub` are different signals, with a connection that sets `SubI` present when `I` is present and `Sub` is active, i.e., `SUSP` is absent. The computation of `pre(SubI)` in `Sub` is subject to suspension: `pre(SubI)` is the last status of `SubI` when `Sub` was active, i.e., the last status of `I` when `SUSP` was absent. Here is an Esterel V7 execution sequence:

```
;
% Output: none
I;
% Output: none
I SUSP;
% Output: X
;
% Output: X Y
I SUSP;
%Output: none
;
%Output: X Y in v5 and X in v7
```

Note that `Y` is absent at last non-SUSP tick, unlike for Esterel V5.

D.5 Multiclock Design

Esterel execution behavior is cycle-based, with the Esterel cycle mapped to the hardware clock cycle in the standard hardware translation. The clock is implicit in each module, and its instants are called ticks and uniformly represented by the `tick` signal. However, the single-clock approach is too limiting for modern circuits, where playing with clocks has become the rule. Several clocks are used to drive several parts of the circuits, and clocks can be gated (i.e., suspended) to save power. The relationship between two given clocks can be quite varied. On one extreme, the clocks can be mutually independent and fully asynchronous; on the other extreme, they can be linked by tight frequency and phase constraints; many intermediate schemes allowing relative frequency changes and variable phase and jitter can also be used.

There are two issues when dealing with multiclock design. On the physical side, one must deal with metastability, which occurs when a register sample its input right at the time a clock front occurs. Metastability is inevitable when the sender and receiver clocks of a signal are asynchronous enough [34]. Special protocol devices called synchronizers are used to resolve it at the price of some latency. On the logical side, assuming metastability has been resolved, designing multiclock circuits becomes much more complex since multiple clock introduce a new level of behavioral asynchrony. Usual hardware description languages give little help in controlling and verifying designs of this type. Fortunately, clean and verifiable multiclock design has been added to Esterel V7 without adding much to its syntax and without changing its semantics. The new **weak suspend** statement is instrumental here. Below, we give a brief explanation of the Esterel V7 multiclock extension.

D.5.1 Clocks and Multiple Units

Esterel V7 introduces a new kind of unit called `multiclock`. This unit can declare standard interface and local signals plus a new kind of signal called a clock. In its body, only few statements are allowed: instantiation of a module explicitly driven by a clock, instantiation of another multiclock unit, clock gating or muxing, and purely combinational equations on signals. A multiclock unit has no clock by itself. Therefore, no **pause** statement and no **pre** expression an appear in its body. Here is an example, where M1 and M2 are conventional modules, with inputs I and outputs O and Y for M1 and input I and output O for M2:

```
multiclock Multi :
  input {C1, C2} : clock;
  input A, B, C;
  output X, Y, Z;
  signal {C3, C4} : clock, S in
    sustain Z <= X or Y   // ok since combinational
  ||
    clock {
      C3 <= C1 if A,       // clock gating
      C4 <= mux(B, C1, C2) // C4 is C1 is B=1 and C2 if B=0
    }
  ||
    run M1 [ clock C3; A / I, S / O, Y / Y ]
  ||
    run M2 [ clock C4; S / I, X / O ]
  end signal
end multiclock
```

Note that the local signal S is generated by M1, which is clocked by C3, and received by M2, which is clocked by C4. Such a signal is called a clock domain crossing signal. Typically, S is registered in M1 to make it electrically clean (glitch-free) and to insert a synchronizer between M1 and M2 if C3 and C4 are asynchronous. Details are outside the scope of this presentation.

The Esterel V7 multiclock design style ensures a clean separation of concerns between the distribution of clocks and signals, performed by multiclock units, and the execution of statements, performed by modules. This is in the GALS (Globally Asynchronous Locally Synchronous) style [66]. Since each module is driven by exactly one clock, there is no need for any change in module definitions. Note that a module has no knowledge of the clock that controls it and simply calls it tick, as in the single clock setting. Note also that each module can gate its clock locally using **weak suspend**.

D.5.2 *Simulation of Multiclock Designs by Single-clocked Designs*

Thanks to the introduction of **weak suspend**, it is easy to give the semantics and software implementation of a multiclock design by translating it into a single-clocked one. The implicit clock of the translated single-clocked design acts as a fictitious simulation clock faster than all actual implementation clocks. This simulation clock does not exist in a hardware implementation.

The translation simply consists in replacing clocks by pure signals and enclosing **run** statement for modules into immediate **weak suspend** statements. No change is needed for module interfaces and bodies. Here is the translation of our example:

```
module Multi :
  input C1, C2;
  input A, B, C;
  output X, Y, Z;
  signal C3, C4, S in
    sustain Z <= X or Y    // sustain ok since combinational
  ||
    sustain {
      C3 <= C1 if A,       // clock gating
      C4 <= mux(B, C1, C2) // C4 is C1 is B=1 and C2 if B=0
    }
  ||
    weak suspend
      run M1 [A / I, S / O, Y / Y ]
    when immediate not C3
  ||
    weak suspend
      run M2 [ S / I, X / O ]
    when immediate not C4
  end signal
end module
```

As far as software implementation is concerned, the translated program can now be compiled by the standard single-clock compiler. A translation to single clock hardware can also be useful to simulate the multiclock circuit using a single-clock FPGA (Field Programmable Gate Array).

Bibliography

Each entry ends with a list of the pages on which it is cited.

[1] Charles André. Representation and analysis of reactive behaviors: A synchronous approach. In *Proceedings of the IEEE-SMC Multiconference CESA 96*, Lille, France, July 1996. Also available as I3S technical report RR 95-02, 1995. ⟨6⟩

[2] Jerry Banks, John S. Carson, Barry L. Nelson, and David M. Nicol. *Discrete-Event System Simulation*. Prentice-Hall, third edition, 2000. ⟨211⟩

[3] Albert Benveniste and Gérard Berry. The synchronous approach to reactive and real-time systems. In *Proceedings of the IEEE*, volume 79(9), pages 1270–1282, 1991. ⟨4⟩

[4] Albert Benveniste, Paul Caspi, Stephen A. Edwards, Nicolas Halbwachs, Paul Le Guernic, and Robert de Simone. The synchronous languages 12 years later. *Proceedings of the IEEE*, 91(1):64–83, January 2003. Invited. ⟨4⟩

[5] G. Berry and L. Cosserat. The synchronous programming language Esterel and its mathematical semantics. In *S. Brookes, G. Winskel, eds., Seminar on Concurrency*, volume LNCS 197, pages 389–448. Springer Verlag, 1984. ⟨10, 42⟩

[6] Gérard Berry. Esterel on hardware. *Philosophical Transactions of the Royal Society of London, Series A*, 19(2):87–152, 1992. ⟨10, 103, 139, 139⟩

[7] Gérard Berry. The constructive semantics of pure Esterel. Draft book available at http://www.esterel-technologies.com/, July 1999. ⟨viii, 6, 10, 11, 15, 21, 21, 21, 29, 41, 42, 44, 44, 48, 48, 50, 103, 104, 104, 105, 119, 176⟩

[8] Gérard Berry. *The Esterel v5 Language Primer*. École des Mines de Paris, CMA and INRIA, Sophia-Antipolis, France, July 2000. www-sop.inria.fr/mejie/esterel/doc/main-papers.html. ⟨viii⟩

[9] Gérard Berry. The foundations of Esterel. In *Proof, Language and Interaction: Essays in Honour of Robin Milner.* MIT Press, 2000. ⟨4⟩

[10] Gerard Berry and Georges Gonthier. The Esterel synchronous programming language: Design, semantics, implementation. *Science of Computer Programming*, 19(2):87–152, 1992. ⟨10, 49⟩

[11] Gérard Berry and Mike Kishinevsky. Extending Esterel for hardware specification and synthesis. Presentation at the Synchron'00 workshop, 2000. ⟨297⟩

[12] Gérard Berry, Sabine Moisan, and Jean-Paul Rigault. Esterel: Towards a synchronous and semantically sound high-level language for real-time applications. In *IEEE Proceedings of the Real-Time Systems Symposium*, 1983. ⟨15⟩

[13] Gérard Berry and the Esterel Team. The Esterel v5.92 system manual. Available at http://www.esterel-technologies.com/, June 2000. ⟨15, 281⟩

[14] Valérie Bertin, Michel Poize, and Jacques Pulou. Une nouvelle méthode de compilation pour le language Esterel [A new method for compiling the Esterel language]. In *Proceedings of GRAISyHM-AAA.*, Lille, France, March 1999. ⟨140⟩

[15] Yves Bertot and Pierre Castéran. *Interactive Theorem Proving and Program Development. Coq'Art: The Calculus of Inductive Constructions.* Springer, 2004. ⟨48⟩

[16] F. Bourdoncle. Efficient chaotic iteration strategies with widenings. In Springer Verlag, editor, *Proceedings of the International Conference of Formal Methods in Programming and Their Applications*, volume LNCS735, 1993. ⟨199⟩

[17] Janusz Brzozowski. Derivatives of regular expressions. *Journal of the ACM*, 11(4):481–494, October 1964. ⟨10⟩

[18] Janusz Brzozowski and Carl-Johan Seger. *Asynchronous Circuits.* Springer-Verlag, 1995. ⟨49⟩

[19] P. Caspi, A. Girault, and D. Pilaud. Distributing reactive systems. In *7th International Conference on Parallel and Distributed Computing Systems, PDCS'94*, Las Vegas, CA, 1994. ⟨9⟩

[20] Massimiliano Chiodo, Daniel Engels, Paolo Giusto, Harry Hsieh, Attila Jurecska, Luciano Lavagno, Kei Suzuki, and Alberto Sangiovanni-Vincentelli. A case study in computer-aided co-design of embedded controllers. *Design Automation for Embedded Systems*, 1(1):51–67, January 1996. ⟨200⟩

[21] Etienne Closse, Michel Poize, Jacques Pulou, Patrick Vernier, and Daniel Weil. Saxo-RT: Interpreting Esterel semantic on a sequential execution structure. *Electronic Notes in Theoretical Computer Science*, 65, 2002. ⟨12, 136, 140, 146⟩

[22] Jean-Louis Colaço, Bruno Pagano, and Marc Pouzet. A conservative extension of synchronous data-flow with state machines. In *Proceedings EMSOFT*, pages 173–182, Jersey City, New Jersey, September 2005. ⟨6⟩

[23] Ron Cytron, Jeanne Ferrante, Barry K. Rosen, Mark N. Wegman, and F. Kenneth Zadeck. Efficiently computing static single assignment form and the control dependence graph. *ACM Transactions on Programming Languages and Systems*, 13(4):451–490, October 1991. ⟨7, 215⟩

[24] B. Dion. Correct-by-construction methods for the development of safety-critical applications. In *Proceedings of the SAE World Congress, paper 04E-129*, 2004. ⟨6⟩

[25] Stephen Edwards. Making cyclic circuits acyclic. In *Proceedings of the Design Automation Conference (DAC'2003)*, Anaheim, California, 2003. ⟨199⟩

[26] Stephen A. Edwards. Compiling Esterel into sequential code. In *Proceedings of the 37th Design Automation Conference*, pages 322–327, Los Angeles, California, June 2000. http://www.dac.com/37proceedings/19_2.pdf ⟨12, 139, 139, 139, 203⟩

[27] Stephen A. Edwards. An Esterel compiler for large control-dominated systems. *IEEE Transactions on Computer-Aided Design of Integrated Circuits and Systems*, 21(2):169–183, February 2002. ⟨12, 136, 139, 146, 185, 211, 212, 229, 229, 229⟩

[28] Stephen A. Edwards, Vimal Kapadia, and Michael Halas. Compiling Esterel into static discrete-event code. In *Proceedings of Synchronous Languages, Applications, and Programming (SLAP)*, Electronic Notes in Theoretical Computer Science, Barcelona, Spain, March 2004. Elsevier Science. ⟨203⟩

[29] Esterel Technologies. *The Esterel v7 Reference Manual*, November 2005. IEEE Standard Proposal. Available at http://www.esterel-technologies.com/. ⟨ix, 15, 297, 298, 309⟩

[30] Jeanne Ferrante, Mary Mace, and Barbara Simons. Generating sequential code from parallel code. In *1988 International Conference on Supercomputing*, pages 582–592, St. Malo, France, July 1988. ACM. ⟨216, 218⟩

[31] Jeanne Ferrante, Karl J. Ottenstein, and Joe D. Warren. The program dependence graph and its use in optimization. *ACM Transactions on Programming Languages and Systems*, 9(3):319–349, July 1987. ⟨203, 215, 216⟩

[32] Xavier Fornari. *Optimisation du contrôle et implantation en circuits de programmes Esterel*. PhD thesis, École des Mines de Paris, CMA, Sophia Antipolis, France, March 1995. ⟨103, 154, 169, 180⟩

[33] Robert S. French, Monica S. Lam, Jeremy R. Levitt, and Kunle Olukotun. A general method for compiling event-driven simulations. In *Proceedings of the 32nd Design Automation Conference*, pages 151–156, San Francisco, California, June 1995. http://suif.stanford.edu/papers/rfrench95.ps ⟨140⟩

[34] Ran Ginosar. Fourteen ways to fool your synchronizer. In *Proceedings ASYNC03*, 2003. ⟨310⟩

[35] Georges Gonthier. *Sémantique et modèles d'exécution des langages réactifs synchrones: application à Esterel*. PhD thesis, Université d'Orsay, Paris, France, March 1988. ⟨10, 42, 51, 139⟩

[36] M. Gordon and T. Melham. *Introduction to HOL: A Theorem Proving Environment for Higher-Order Logic*. Cambridge University Press, 1993. ⟨48⟩

[37] Gary Hachtel and Fabio Somenzi. *Logic Synthesis and Verification Algoritms*. Kluwer Academic Publishers, 1996. ⟨169⟩

[38] N. Halbwachs, P. Caspi, P. Raymond, and D. Pilaud. The synchronous dataflow programming language Lustre. In *Proceedings of the IEEE*, volume 79(9), pages 1305–1320, 1991. ⟨6, 42, 297, 305⟩

[39] Nicolas Halbwachs. *Synchronous Programming of Reactive Systems*. Kluwer academic Publishers, 1993. ⟨3, 4⟩

[40] David Harel. Statecharts: A visual formalism for complex systems. *Science of Computer Programming*, 8, 1987. ⟨3, 7⟩

[41] David Harel and Amir Pnueli. On the development of reactive systems. In *K. Apt (ed.), Logics and Models of Concurrent Systems*, NATO ASI, pages 477–498, New York, 1985. Springer-Verlag. ⟨3⟩

[42] Reinhold Heckmann, Marc Langenbach, Stephan Thesing, and Reinhard Wilhelm. The influence of processor architecture on the design and the results of WCET tools. *Proceedings of the IEEE*, 91(7), July 2003. ⟨6⟩

[43] Arend Heyting. *Intuitionism: An Introduction*. North-Holland Publishing, Amsterdam, 1971. Third Revised Edition. ⟨44⟩

BIBLIOGRAPHY 319

[44] IEEE, 345 East 47th Street, New York, NY 10017-2394, USA. *IEEE Std p1364-2001, IEEE Standard Hardware Description Language Based on the Verilog©Hardware Description Language*, 2001. ⟨7⟩

[45] Hamoudi Kalla, Jean-Pierre Talpin, David Berner, and Loic Besnard. Automated translation of C/C++ models into a synchronous formalism. In *Proceedings of the IEEE International Conference on the Engineering of Computer-Based Systems (ECBS)*, Potsdam, Germany, 2006. ⟨7⟩

[46] P. LeGuernic, T. Gauthier, M. LeBorgne, and C. LeMaire. Programming real-time applications with Signal. In *Proceedings of the IEEE*, volume 79(9), pages 1321–1336, 1991. ⟨6⟩

[47] Jan Lukoschus. *Removing Cycles in Esterel Programs*. PhD thesis, Christian-Albrechts-Universität Kiel, Department of Computer Science, 2006. ⟨199⟩

[48] Sharad Malik. Analysis of cyclic combinational circuits. In *Conference on Computer Aided Design (ICCAD)*, pages 618–625, Santa Clara, CA, USA, November 1993. IEEE Computer Society. ⟨11⟩

[49] Florence Maranichi. The Argos language: Graphical representation of automata and description of reactive systems. In *IEEE Workshop on Visual Languages*, Kobe, Japan, 1991. ⟨6⟩

[50] Peter M. Maurer. Event driven simulation without loops or conditionals. In *Proceedings of the IEEE/ACM International Conference on Computer Aided Design (ICCAD)*, pages 23–26, San Jose, California, November 2000. ⟨203⟩

[51] Stanley Mazor and Patricia Langstraat. *A Guide to VHDL*. Kluwer, 1992. ⟨7, 7⟩

[52] R. Milner. *Communication and Concurrency*. Series in Computer Science. Prentice Hall, 1989. ⟨55⟩

[53] Osama Neiroukh, Stephen Edwards, and Xiaoyu Song. An efficient algorithm for the analysis of cyclic circuits. In *Proceedings of the Symposium on VLSI (ISVLSI)*, Karlsruhe, Germany, 2006. ⟨199⟩

[54] Gordon Plotkin. A structural approach to operational semantics. Technical Report report DAIMI FN-19, University of Aarhus, 1981. ⟨48, 55⟩

[55] Dumitru Potop-Butucaru. Fast redundancy elimination using high-level structural information from Esterel. RR 4330, INRIA, 2001. ⟨169⟩

[56] Dumitru Potop-Butucaru. *Optimizations for Faster Simulation of Esterel Programs*. PhD thesis, École des Mines de Paris, CMA, Paris, France, November 2002. ⟨12, 49, 50, 139⟩

[57] Dumitru Potop-Butucaru. Optimizations for faster execution of Esterel programs. In *Proceedings of the 1st International Conference on Formal Methods and Models for Codesign (MEMOCODE)*, pages 227–236, Mont St. Michel, France, June 2003. ⟨12⟩

[58] Dumitru Potop-Butucaru, Robert de Simone, and Jean-Pierre Talpin. The synchronous hypothesis and synchronous languages. In R. Zurawski, editor, *Embedded Systems Handbook*. CRC Press, 2005. ⟨4⟩

[59] K. Schneider. A verified hardware synthesis for Esterel. In *Distributed and parallel embedded systems (DIPES)*, pages 205–214, 2000. ⟨48⟩

[60] K. Schneider. Embedding immperative synchronous languages in interactive theorem provers. In *Proc. Conference on application of concurrency to systems design (ACSD)*, pages 143–156, 2001. ⟨48⟩

[61] K. Schneider. Proving the equivalence of microstep and macrostep semantics. In *15th International Conference on Theorem Proving in Higher Order Logics*, 2002. ⟨47⟩

[62] K. Schneider and M. Wenz. A new method for compiling schizophrenic synchronous programs. In *Proceedings CASES2001*, Atlanta, Georgia, USA, 2001. ⟨6, 47, 307⟩

[63] Ellen Sentovich, Horia Toma, and Gérard Berry. Efficient latch optimization using exclusive sets. In *Proceedings DAC'97*, 1997. ⟨11, 109, 169⟩

[64] E.M. Sentovich, K.J. Singh, L. Lavagno, C. Moon, R. Murgai, A. Saldanha, H. Savoj, P.R. Stephan, R.K. Brayton, and A.L. Sangiovanni-Vincentelli. SIS: A system for sequential circuit synthesis. Memorandum UCB/ERL M92/41, University of California at Berkeley, 1992. ⟨169⟩

[65] Thomas Shiple, Gérard Berry, and Hervé Touati. Constructive analysis of cyclic circuits. In *Proccedings EDTC*, Paris, France, 1996. ⟨11, 42, 46, 49⟩

[66] Sandeep K. Shukla and Michael Theobald. Special issue on formal methods for globally asynchronous and locally synchronous (GALS) systems. guest editorial. *Formal Methods in System Design*, 28(2), March 2006. ⟨311⟩

[67] Barbara Simons and Jeanne Ferrante. An efficient algorithm for constructing a control flow graph for parallel code. Technical Report TR-03.465, IBM, Santa Teresa Laboratory, San Jose, California, February 1993. ⟨215, 216, 228⟩

[68] Bjarne Steensgaard. Sequentializing program dependence graphs for irreducible programs. Technical Report MSR-TR-93-14, Microsoft, October 1993. ftp://ftp.research.microsoft.com/pub/TR/TR-93-14.ps ⟨216, 228⟩

[69] Horia Toma. *Analyse constructive et optimisation séquentielle des circuits générés à partir du langage synchrone réactif Esterel.* PhD thesis, École des Mines de Paris, CMA, Paris, France, September 1997. ⟨11, 169, 180, 193, 199, 199⟩

[70] Hervé Touati and Gérard Berry. Optimized controller synthesis using Esterel. In *Proceedings of the International Workshop on Logic Synthesis IWLS'93*, Lake Tahoe, USA, 1993. ⟨169⟩

[71] Anne Sjerp Troelstra and Dirk van Dalen. *Constructivism in Mathematics. An Introduction.* North-Holland, 1988. ⟨44⟩

[72] Daniel Weil, Valérie Bertin, Etienne Closse, Michel Poize, Patrick Venier, and Jacques Pulou. Efficient compilation of Esterel for real-time embedded systems. In *International Conference on Compilers, Architecture, and Synthesis for Embedded Systems (CASES)*, pages 2–8, San Jose, California, November 2000. ⟨12, 140, 200⟩

[73] Jia Zeng, Cristian Soviani, and Stephen A. Edwards. Generating fast code from concurrent program dependence graphs. In *Proceedings of Languages, Compilers, and Tools for Embedded Systems (LCTES)*, pages 175–181, Washington, DC, June 2004. ⟨215⟩

Index

*, 257
+, 257
-, 257
/, 257, 270
:=, 261, 271
;, 16, 260
<, 257
<=, 257
<>, 257
=, 257
=>, 277
>, 257
>=, 257
?, 19, 256, 258, 302
??, 256, 258
[, 261
#, 277
%, 253
#define, 281, 283, 284, 286, 287
], 261

abort, 18, 254, 261, 274
abort
    multi-way, 262
ABRO example, 7, 104
actions, 150
acyclicity, 14, 47, 180–185
    and optimization, 180
    compiler-independent, 180
analysis of potentials, 56, 58
and, 235, 243, 254, 257, 259
Argos, 6
argument
    of string type, 287
    reference, 287, 295
    value, 287

arrays, 297, 298
    slicing, 298
assert, 306
assertions, 306
    implicit, 307
assignment
    #define, 284
    full copy, 284
    function, 284
    statement, 261
    string, 283
asynchronous environment, 13
asynchrony, 310
atomic reaction, 6, 282, 291
await, 17, 18, 254, 262, 263, 265

BDD, 46
behavioral transition, 55
bin2u, 300
bit vectors, 299
    conversion, 300
bool, 298
Boolean
    circuit, 45, 104
    Decision Diagram, *see* BDD
    register, 104, 106
boolean, 33, 257, 283, 298
boot register, 114
boot statement, 56
$\perp$ (bottom), 45, 46, 51, 58, 60, 64, 80–94, 97–99, 101, 102, 129, 130
bullet, 80–83, 100

C interface, 281–296
    #define, 281, 283, 286, 287

INDEX

assignment function, 284
atomic reaction, 282
atomicity, 291
`boolean`, 283
combined input, 288, 289
constant, 281, 286
data-handling, 281
`double`, 283
equality function, 285
execution example, 282
execution shell, 12, 247, 281, 282, 294
extern constant, 286
extern definition, 283
extern function, 286
extern procedure, 287
files, 283
`float`, 283
function, 281, 286
function names, 282
inequality function, 285
initialization, 290
inline assignment, 284
inline constant, 286
inline function, 286
inline procedure, 287
input, 289
input event, 282, 288
input function, 281, 282, 288
input order, 288
inputoutput, 289
`integer`, 283
interrupt, 291
no reentrance, 291
output, 289
output event, 282
output function, 281, 282, 289
output order, 289
predefined types, 283
procedure, 281, 287
reaction, 290
reaction function, 281, 282, 288
relation, 291
reset, 290

return signal, 289
sensor function, 282, 290
simultaneity, 288
`strcpy`, 283
`string`, 283
`string` assignment, 283
`STRLEN`, 283
`struct`, 284
task, 281, *see* Task
type, 281, 283
`typedef`, 284
Côte d'Azur, vii
`call`, 254, 257, 264
`call:`, 156
Can, 48, 49, 56, 58–68, 79, 80, 84, 85, 90, 92, 95–102, 121, 172
Can^+, 48, 49, 56, 58–68, 71, 80, 102
`case`, 254, 265, 268
causal, 25
causality, 6
    constructive, 46, 106
    cycle, 8, 19, 26, 42
    error, 28
    loop, 45
causally
    incorrect, 42
CBS, *see* constructive behavioral semantics
CEC, *see* Columbia Esterel Compiler
circuit
    code, 11, 135, 136
    selection, 114
    with data, 103, 104, 107, 109, 112
`clock`, 311
code
    C interface, *see* C interface
    circuit, 10, 11, 135, 136
    control-flow, 12, 135, 136
    duplication, 218
    FSM, 9, 135
    generation, 9–12
code generation

for acyclic GRC, 185–193
for cyclic specifications, 193–199
from GRC, 179–201
from the PDG, 215–229
in the Columbia Esterel Compiler, 203–233
structured, 212
using lists, 203–215
`code2u`, 300
Columbia Esterel Compiler, 142, 203–233
combinational gate, 104
`combine`, 254
combine function, 19
comments, 253
Compilation Techniques
classes, 135
compiler
classes of, 135
Columbia Esterel, *see* Columbia Esterel Compiler
Esterel Studio, 10, 47, 50, 137, 145, 159, 240, 247
Esterel V2, 9, 10, 13, 136
Esterel V3, 9, 10, 43, 51, 136
Esterel V4, 136
Esterel V4–V6, 10
Esterel V5, ix, 12, 33, 37, 136, 137, 235, 240, 255, 273, 297–299, 301, 305, 309, 310
Esterel V5_92, 155, 197, 201
Esterel V7, vii, ix, 4, 12, 43, 50, 110, 137, 145, 150, 159, 169, 175, 240, 245, 247, 256, 269, 297, 299, 301, 304–307, 309–312
INRIA, 9, 37, 50, 118, 137–139, 145, 185, 236
Saxo-RT, 47, 140–143, 145, 146, 169, 176, 179, 180, 201–203, 211, 230, 231
Synopsys, 139–142, 145, 146
completion codes, 51, 55, 73, 117
confluence, 100

connection code, 76, 100
constant, 256, 281
`#define`, 281, 283, 286
atoms, 258
definition, 286
example, 286
extern, 283, 286
`constant`, 254, 270, 277
constructive, 100
approach, 43
behavioral semantics, 47–49, 55–77, 79–81, 84, 85, 90, 95, 97, 100, 101
causality, 104, 106
circuit model, 103
circuit semantics, 47, 49–50, 103–131
circuit translation, 45
circuit with data, 107, 112
operational semantics, 49, 79–102
semantics, 42
value propagation, 105
constructiveness, 195
control arcs, 150
control dependence, 141
control-flow code, 14, 135, 136
conventions, 50
`copymodule`, 34, 254, 270
Coq, 48
correctness, 50
COS, *see* constructive operational semantics
counted delay, 242
counter, 242
cycle-based execution, 5
cyclic arbiter, 27
cyclicity, 15, 26
data
abstraction, 109
action, 103, 107–109, 126
dependency, 108, 128
encoding, 129
`data`, 300
data arc, 149, 150, 167, 219–221

326   INDEX

data handling, *see* C interface
de Simone, Robert, vii
declaration, 32
  extern, 286
  interface, 302
  shared variable, 165
  signal, 271
  trap, 273
  variable, 16, 273
decorated terms, 56, 57, 60, 63, 64
DEI, *see* FEIF
delay
  counted, 242
  expression, 241, 261
  expressions, 259
demit:, 156
dependence
  control, 141
  factoring, 150
depth, 113–121, 123, 124, 127, 128, 131
depth-first search, *see* DFS
determinism, 72, 100, 103
DFS, 219, 221, 224
digital synchronous circuit, 6, 7
discrete time, 5
do, 254, 262, 263, 265, 273, 306
do-upto (deprecated), 264
do-watching (deprecated), 264
dodown, 306
dominators, 212
dopar, 304
double, 33, 257, 283, 298
doup, 306
dual-rail encoding, 198
dumb:, 156

each, 254, 265
École des Mines de Paris, 137
EEC, 228
else, 254, 266, 268
elsif, 254, 266
emit, 17–19, 22, 41, 43, 44, 56, 57, 61, 68, 72–74, 80, 81, 90, 97, 160, 254, 256, 264, 297, 302, 305, 306
encoding
  dual-rail, 198
encoding, state, *see* state encoding
end, 254
enter primitive, 148, 149, 190, 191
equality function, 285
Esterel
  history, 136–137
  introduction to, 15–37
  kernel, 20–24, 51, 57, 75, 101, 128, 235, 237, 253
  pure, 21, 22, 24, 85, 101, 103, 111
  reference manual, 253–279
  region, vii
  semantics of, 21, 41–131
  Studio, 10, 47, 50, 137, 145, 159, 240, 247
  V2, 9, 10, 13, 136
  V3, 9, 10, 43, 51, 136
  V4, 136
  V4–V6, 10
  V5, ix, 12, 33, 37, 136, 137, 235, 240, 255, 273, 297–299, 301, 305, 309, 310
  V5_92, 155, 197, 201
  V7, vii, ix, 4, 12, 43, 50, 110, 137, 145, 150, 159, 169, 175, 240, 245, 247, 256, 269, 297–313
Esterel Technologies, ix, 137, 297
evaluation order, 106
event, 281, 282
every, 254, 265
every-do, 265
example
  ABRO, 7
  Arbiter, 27
  await expansion, 18
  C execution, 282
  causality, 26
  combined input, 289

completion code, 53
constant, 286
constructive, 43
FIFO, 35–36
function, 286
illustrating cyclic, 181
input, 288
MainExample, 24
Monster, 26
multiple reincarnation, 29
preemption, 24
procedure, 287
reaction, 288
reincarnation, 29
Small, 114
type, 285
`exclusive:`, 154, 170, 172
`exec`, 247, 254, 257, 265
`ExecStatus`, *see* Task
execution, *see* C interface
execution shell, 13
`exit`, 16, 22, 52, 54, 61, 68, 73, 74, 89, 97, 124, 164, 254, 256, 266
exit primitive, 148, 149, 190, 191
experiment
    code size, 202
    execution speed, 230, 233
    execution time, 202
    parallel branch redundancy, 171, 172
    state encoding, 190
expressions, 257–259
    combinational, 236–238
    data, 258
    delay, 241, 259, 261, 272
    signal, 235–242, 258
    trap, 242, 243, 259
`extends`, 300, 303, 304
external
    constant, 286
    function, 286
    procedure, 287
    type, 298

external edge condition, *see* EEC

`false`, 33, 254
FEIF, *see* DEI
Field-Programmable Gate Array, *see* FPGA
fields
    named, 299
    overlapping, 300
file
    `PROG.c`, 281, 283
    `PROG.h`, 281, 283
    `PROG.strl`, 281
    `PROG_data.c`, 281, 283
`finalize`, 245–247
fixpoint, 106
`float`, 33, 257, 283, 298
`for`, 304, 306
FPGA, 312
FSM
    expansion, 9
FSM code, 135
function, 257, 281
    `#define`, 281, 283, 286
    assignment, 284
    Can, 48, 49, 56, 58–68, 79, 80, 84, 85, 90, 92, 95–102, 121, 172
    Can$^+$, 48, 49, 56, 58–68, 71, 80, 102
    combine, 19, 290
    definition, 286
    equality, 285
    example, 286
    extern, 283, 286
    in expressions, 258
    inequality, 285
    inline, 286
    input, 281, 282, 288, 289
    kill task, 295
    Must, 48, 49, 56, 58–68, 71, 80, 95, 102
    names, 282
    output, 281, 282, 289

potential, *see* potential function
reaction, 135, 139, 141, 148, 155, 194, 281, 282, 288, 290
reset, 290
resume task, 296
return signal, 289
sensor, 282, 290
start task, 295
suspend task, 295
`function`, 254, 270, 278

GALS, 311
generated code, *see* C interface
`generic`, 299
generic types, 299
Globally Asynchronous Locally Synchronous (GALS), 311
`gray2u`, 300
GRC format, 145–177
    computation nodes, 149
    control-flow graph, 207–208
    flowgraph, 145
    in CEC, 204–211
    node interface, 150
    operators, 146
    optimization, 146
    selection tree, 145–148, 153–156, 158, 167, 170, 172–174, 185, 186, 188, 189, 191, 193, 195, 206, 207
guard variable, 215, 216, 218, 221, 228
    fusion, 228
guessing, 105

`halt`, 254, 266
`handle`, 242, 254, 273
history, 136–137
HOL, 48
`host`, 298, 299
host language, 13, 15, 31
    data types, 297
    interface, 13, 31

identifiers, 253
`if`, 23, 94, 96, 99, 128, 163, 254, 266, 304
`immediate`, 241, 254, 261, 263
`in`, 254
`index`, 305
inequality function, 285
`init:`, 156
inline
    constant, 286
    function, 286
    procedure, 287
input
    event, 13, 55, 281, 282
    function, 281, 282, 288
    function (example), 288
    order, 288
`input`, 254, 276
input/output buffering, 13
inputoutput
    input function, 289
    output function, 289
`inputoutput`, 254, 276
INRIA compiler, ix, 137–139, 211
`integer`, 33, 257, 283
integer ranges, 298
interactive system, 3
interface, 302
    mirroring, 304
`interface`, 302, 304
interrupt, 291
intuitionistic constructive logic, 44

kernel Esterel, 15, 20, 21, 51, 57, 75, 101, 128, 235, 237, 253
    restriction to, 51
kernel expansion
    of concurrent traps, 243
    of general trap, 245
    of `pre`, 238
    signal expressions, 237
keywords, 253, 254

La Londes les Maures, vii
LC format, 137, 139, 200

LCA, 217
least common ancestor, *see* LCA
levelizing, 211
linked list, 203, 204, 211–213, 231
literals, 33, 253–255
    Boolean, 255
    numeric, 254
    string, 255
logic
    combinational, 104
    gate, 104
    intuitionistic constructive, 44
    three-valued, 46
logically correct, 44
loop
    instantaneous, 73
    preempted, 25
    safety, 73, 74
loop, 16, 22, 57, 62, 66, 70, 73, 74,
    83, 87, 88, 97, 98, 118,
    160, 254, 265, 267
loop-each, 267
Lustre, 6, 42, 297, 305

macro-step, 55
map, 299
mathematical foundation, 41
metastability, 310
microsteps, 47, 49, 79–83, 87, 88,
    90, 92, 93, 97, 99, 101,
    113
mirror, 304
mod, 254, 257
module, 15, 31
    instantiation, 34, 36, 270, 309
module, 31, 254, 275
multi-way await, 263
multiclock, 297
multiclock, 310–313
multiple exec, 251
Must, 48, 49, 56, 58–68, 71, 80, 95,
    102

namespaces, 255
next, 297, 302

non-constructive, 100, 111
non-deterministic, 100
non-interruptible reaction, 6
nondeterminism, 72, 100, 103, 297
nonterm:, 155, 170–172, 174, 177,
    186, 187, 190, 192
not, 235, 243, 254, 257, 259
nothing, 16, 22, 52, 61, 68, 73,
    74, 85, 117, 160, 162, 254,
    268

onehot2u, 300
operator, 33, 253
    *, 257
    +, 257
    -, 257
    /, 257, 270
    <, 257
    <=, 257
    <>, 257
    =, 257
    =>, 277
    >, 257
    >=, 257
    ?, 19, 256, 258, 302
    ??, 256, 258
    #, 277
    and, 235, 254, 257, 259
    mod, 254, 257
    not, 235, 254, 257, 259
    or, 235, 254, 257, 259
    pre, 235, 238, 239, 258
    precedence, 258
    predefined, 257, 258
or, 235, 243, 254, 257, 259
oracle, 302
output
    event, 282
    function, 281, 282, 289
    order, 289
output, 254, 276

parallel, 52, 161, 172, 190
    associativity, 261
    behavior, 9

330 INDEX

branch redundancy, 171, 172
children, 192
communication, 17, 19
completion codes, 52, 120, 152
construct, 16
cyclic dependencies, 183
data actions, 108
decorations, 83
depth circuit, 120
example, 8, 10, 20, 28, 53, 171, 192
false dependencies, 175
fine-grained, 24
for-dopar, 305
fork node, 152
loop safety and, 75
node, 147, 148, 154, 172
non-deterministic interleaving, 100
non-terminating branches, 171, 172
potential functions, 96
preemption, 171
reincarnation of assert, 307
resuming, 66, 87
rules for, 86
scheduling, 137
selection, 58, 67
selection node, 152
selection tree, 146, 147, 154
semantic rules for, 71
signal potential, 62
simplification, 192
state encoding, 187, 191
statement, 16, 119
status, 83, 139, 190
structure, 170, 174
surface circuit, 120
sync node, 157, 174
sync primitive, 149
synchronization, 8, 51, 54, 63, 87, 104, 139, 146, 152
synchronizer, 120, 121, 157, 167, 180, 183, 190

splitting, 183–185
synchronizer refinement, 184
syntax, 260
traps, 243
`parallel:`, 154, 170, 172, 190
parameter
by-reference, 278
by-value, 278
`pause`, 17, 22, 41, 52, 53, 56–58, 61, 63, 64, 69, 73, 74, 80–82, 85, 97, 104, 106, 107, 113, 114, 117, 148, 149, 159, 254, 263, 268
PDG, 203, 215–219, 221, 222, 224, 226, 228, 233
code generation from, 216
cutting, 216
data dependencies, 216
definition, 216–218
node duplication, 224–225
nodes, 217
restructuring, 215, 222–224
scheduling, 216, 218–228
semantics, 217
structure, 217
PLCA, 217, 221, 224
`positive`, 254, 269
potential function, 48, 60, 95
Can, 79, 90, 92
Can^+, 48, 59, 60
CBS, 55, 101
characterization, 102
code pruning, 102
completion codes, 73
control-flow, 48
COS, 80, 101
COS vs. CBS, 101
cycle, 63
data, 99
definition, 60
derived statements, 96
deterministic, 72
inductive definitions, 68
loop, 62, 97

Must, 49, 68
notation, 59
parallel, 67, 96
parallel synchronizer, 120
pause, 64
sequence, 62
sequencing, 96
shared variables, 84
signal, 62
signal absence, 90
signal declaration, 97
suspend, 65
trap, 67
triplets, 96
potentials
 analysis of, 56, 58
pre, 235, 238, 239, 258, 297
pre
 and signal expressions, 240
 and valued signals, 239
predefined type, 283
predefined types, 33
predicate least common ancestor, *see* PLCA
preemption, 25
 with suspend, 272
present, 17, 22, 57, 61, 64, 65, 69, 73, 74, 83, 91, 92, 96, 98, 123, 163, 254, 263, 268, 304
procedure, 257, 281
 #define, 281, 283, 287
 call, 264
 definition, 287
 example, 287
 extern, 283, 287
 inline, 287
 reference argument, 287
 value argument, 287
procedure, 254, 270, 278
PROG, 281
program dependence graph, *see* PDG
proof, 100
 equivalence, 48

example, 72
intuitionistic, 44
of determinism, 72
pure Esterel, 21, 22, 24, 85, 101, 103, 111

Quartz, 47, 48

reaction, 5, 135, 136, 139, 231
 active code in, 12, 49
 atomic, 6
 behavioral transition, 79
 bullet, 80
 causality, 6, 25
 CBS, 48, 55
 combinational gates for, 104, 110
 completion status, 83
 computing, 8, 45, 55
 constructive circuit semantics, 103
 constructiveness, 111
 control propagation in, 81
 control-flow, 16, 80
 execution in the next, 57
 execution shell and, 13
 first, 114
 function, 12, 13, 53, 281, 282, 288
  example, 288
 function atomicity, 282
 host language, 31
 inputs at beginning, 105
 microsteps, 49, 79, 99
 non-interruptible, 6
 output of, 8
 overlapping, 13
 pause and, 17, 56, 106
 predictable, 6
 register values, 106
 restarting, 83
 semantics of, 49
 shared variable, 84
 signal emission in, 19
 signal propagation, 47

signal status in, 43
START wire, 107
static scheduling and, 14
surface and depth, 115
ternary simulation, 50
transition function, 64
variable values across, 84
zero-time, 5, 23
reactive system, 3
reductio ad absurdum, 44
`reg`, 301
Register Transfer Level, *see* RTL
reincarnation, 15, 29–31, 113, 211
and tasks, 250
multiple, 29
`relation`, 254
relations, 34, 277
`repeat`, 254, 263, 269
reserved words, 254
reset, 290
`return`, 34, 254, 265, 276
return function, 289
RTL, 7
`run`, 34, 36, 254, 270
run-time, *see* C interface

Safety-Critical Application Development Environment, *see* SCADE
Saxo-RT, 47, 140–143, 145, 146, 169, 176, 179, 180, 201–203, 211, 230, 231
Saxo-RT compiler, 140–142
SCADE, 6
SCC, 176, 179, 195–197
code generation from, 197–199
scheduling, 137, 140–142, 146, 159, 179, 185, 190, 212
clusters, 211, 212
constraints, 204
dynamic, 179, 180, 195
NP-completeness, 211
of GRC, 192–193
overhead, 212
PDG, 218–228

priority queue, 211
static, 179, 180, 185, 195
under forks, 215
schizophrenia, 15, 29–31
Schneider, Klaus, 47
Scott domain, 46
selection circuit, 114
semantics
circuit, 104
circuit vs. COS, 128–131
constructive, 42, 104
constructive behavioral, 47–49, 55–77, 79–81, 84, 85, 90, 95, 97, 100, 101
constructive circuit, 47, 49–50, 103–131
constructive operational, 49, 79–102
COS vs. CBS, 101–102
flavors, 47
notation, 54, 59
structural operational, 55
`sensor`, 34, 254, 276
sensor function, 282, 290
sensors, 256, 301
separators, 253
`seq`, 306
`shared`, 23, 24, 83, 92, 93, 97, 99, 165
shared variable, 92–93
assignment, 84
Can, 95
connection code, 100
declaration, 165
dependency, 128
encoding, 126
initialization, 127
interface, 100, 128
non-constructive, 100
ordering, 108
potential function, 84
potentials, 95
restriction operator, 95
synchronization, 79, 80, 84

transformations, 76
updates, 156
signal, 17–20, 32, 34, 256, 276
    arrays, 306
    dependencies, 150
    equational form, 305
    event, 51
    expressions, 235–242, 258
    expressions and pre, 240
    hubs, 150
    initialization, 302
    link set, 158–165, 167
    oracle, 302
    pure, 17, 19, 256
    registered, 301
    relation, 34
    simultaneous, 288
    temporary, 301
    tick, 34
    value-only, 301
    valued, 17–20, 256
signal, 17, 22, 51, 57, 63, 64, 67, 71, 73, 74, 83, 89, 90, 97, 98, 122, 162, 254, 256, 270, 271
Signal (language), 6
signed, 298
simplification, 153, 169, 170, 173–176, 183, 193, 195, 200
simulation
    constructive, 195
    ternary, 45, 46
SLAP workshop, 203
slicing, array, 298
speculation, 44
SSA, 7
state decoding, 149, 185
    in circuits, 186
state encoding, 139, 145, 182, 185–190
    complexity, 189
    in circuits, 186
state update, 149
Statecharts, 7

statement
    boot, 56
statements, 260–275
static scheduling, 14
Static Single Assignment, *see* SSA
STD_EXEC, 296
strcpy, 283
string, 33, 257, 298
string, 283
    as procedure argument, 287
    assignment, 283
    strcpy, 283
    STRLEN, 283
STRLEN, 283
strongly-connected component, *see* SCC
structural operational semantics, 55
surface, 113–121, 123, 124, 126–131
suspend, 17, 22, 57, 61, 65, 69, 73, 74, 82, 83, 91, 96, 98, 124, 164, 254, 263, 272
suspension
    weak, 307
sustain, 254, 273, 297, 305, 306
switch primitive, 148, 190
sync primitive, 148, 190
SyncCharts, 6
SYNCHRON, vii
synchronization, 19
synchronizer, *see* parallel synchronizer
synchronous
    circuit, 6
    hypothesis, 4
    language, 6
Synopsys compiler, 139–140, 146, 203

task, 34, 247–251, 257, 265, 281
    active field, 293
    exec_index field, 294
    execStatus, 292
    functional interface, 295
    interface, 292

kill field, 293
kill function, 295
low-level interface, 292
multiple instances, 251
preemption, 250
pRet field, 294
prev_active field, 293
prev_suspended field, 293
pStart field, 294
reference argument, 295
reincarnation, 294
resume function, 296
return signal, 250
start field, 293
start function, 295
STD_EXEC macro, 296
suspend function, 295
suspended field, 293
suspension, 247
task_exec_index field, 294
task, 254, 270, 278
temp, 301
terms
    decorated, 56, 57, 60, 63, 64
ternary circuit simulation, 45, 46
test primitive, 148, 190
then, 254, 266
theoretical foundation, 44
thesis
    Fornari's, 169, 180
    Gonthier's, 42, 51
    Lukoschus's, 199
    Potop-Butucaru's, vii
    Toma's, 169, 180, 193, 199
tick, 34, 256, 310
timeout, 254
times, 254, 263, 269
tokens, 253
topological sort, 47
transformational system, 3
transition
    behavioral, 55
    system, 56
transition rules, 49

microstep, 47, 49, 79–83, 87, 88, 90, 92, 93, 97, 99, 101, 113
translation
    constructive circuit, 45
trap, 242–245, 256
    concurrent, 243–244
    encoding, 51
    expressions, 243, 259
    handler, 242
    pure, 256
    valued, 242, 244, 256, 258
trap, 16, 22, 57, 63, 67, 70, 71, 73, 74, 83, 88, 89, 97, 98, 125, 165, 242, 254, 273
true, 33, 254
type, 32, 257, 281
    #define, 284
    assignment function, 284
    boolean, 283
    checking, 36, 257, 258
    definition, 281, 283
    double, 283
    equality function, 285
    example, 285
    float, 283
    generic, 299
    inequality function, 285
    inline assignment, 284
    integer, 283
    of signal, 271
    predefined, 33, 257, 283
    string, 283
    struct, 284
    typedef, 284
    user-defined, 257, 283
type, 254, 270, 277, 298, 299

u2bin, 300
u2code, 300
u2gray, 300
u2onehot, 300
underscores in identifiers, 253
unsigned, 298
upto, 254

user-defined type, *see* C interface

V2–V7, *see* Esterel V2–V7
V7
    interface, 302
    modules, 309
    multiclock, 310
    oracles, 302
`value`, 301
`var`, 16, 24, 83, 93, 94, 97, 99, 165, 254, 273
variable, 16, 93–94, 256
    guard, *see* guard variable
    initialization, 16
    sequential, 93–94
    shared, 76, 79, 80, 84, 92–93, 95

verification assertions, 306
VHDL, 7
`void:`, 155, 170–174, 187

`watching`, 254
WCET analysis, 13
weak
    abort, 274
    suspend, 297
`weak`, 254, 274
`weak suspend`, 307
`when`, 254, 261, 263, 272
whitespace, 253
`with`, 254
Worst-Case Execution Time (WCET), 13

Printed in the United States
77239LV00002B/1-45